Professional Ethics for Research and Development Activities

Dag Slotfeldt-Ellingsen

Professional Ethics for Research and Development Activities

 Springer

Dag Slotfeldt-Ellingsen
Oslo, Norway

ISBN 978-3-031-25483-3   ISBN 978-3-031-25484-0   (eBook)
https://doi.org/10.1007/978-3-031-25484-0

This Springer imprint is published by the registered company Springer Nature Switzerland AG
The registered company address is: Gewerbestrasse 11, 6330 Cham, Switzerland

Preface

In this book, I address ethical issues that researchers encounter in everyday life.

Much of what has been written about research ethics focuses on the development of and justification for research ethics, and on the grey zones between right and wrong. Much is also about poor research morality, focusing on researchers who have committed sensational scientific misconduct. Philosophizing on the nature of research and the ethos of the researcher can provide a deeper understanding of ethics and illuminate difficult ethical dilemmas, and much can be learned from the mistakes of other researchers. But for the majority of researchers, research ethics is primarily a question of finding *quick, simple and direct answers to what, in different situations, would surely be considered responsible research practices*. To help each researcher in this respect, both national and international written codes of conduct for research have been established, mainly from the 1990s onwards.

This book is an introduction to professional ethics for research and development activities and ethically responsible research practice in a research organization. It is relevant for all areas of research. Morality, however, is not something one can simply 'learn'. Therefore, the book is written with a different fundamental tone than regular textbooks so that it might both *make the reader aware* of how morality plays a role in various daily tasks one has in a research organization and *convey knowledge and experience material* about society's and the research community's view of how different ethical issues should be solved. From this, the readers will acquire the basics they need to know, understand and be aware of in order to be ethically qualified as employees of a research organization, and to be able to deal with common ethical issues associated with research.

The primary target group for the book is therefore students and new employees in research organizations, newly appointed leaders and employees in administrative positions that have something to do with research and everyone elected to committees related to ethics. The courses in research ethics that everyone at educational institutions must attend often consist only of a few lectures that cover the subject's core. The book is appropriate support for such courses in that it is easy to read and suited to self-study. It also covers a wide range of topics, and students can choose to focus on what in practice will be most relevant to them in their future professional undertakings. For research organizations, the book can be used as a means of ensuring that all employees are aware of their ethical responsibilities.

Both experience and surveys among researchers suggest that a surprising number of established researchers possess inadequate knowledge of national and international codes of conduct for research. The book is therefore also designed so that experienced researchers, leaders and administration personnel can use it to support their own assessments when faced with unfamiliar ethical issues.

Emphasis has been placed on treating each topic so that students and others without research experience will understand the issues. The book is also written so that one can read about a single topic of interest, and be informed about it, without having to read the whole book. The book should therefore be of help to most people involved in research in different ways – be it at universities, colleges, research institutes, research units in companies and organizations, scientific publishers or in public research administration.

In addition, the book stands out in particular in four ways:

1. The professional ethics of employees in a research organization includes not only ethical issues that are specific to the research work itself but also more general ethical issues. An example is issues related to employee morale in general, which are found in all professions. Here, however, I focus on what is particularly relevant in research organizations. The book, therefore, also deals with issues that go beyond what is usually discussed in textbooks and articles on research ethics.
2. The relationship between the individual professional and society is central to all professional ethics; so too in research ethics, where in a number of contexts the emphasis on researchers' responsibility to society seems to be growing. This book joins this trend but goes further by emphasizing that to a great extent research ethics are researchers' response to society's expectations towards research. I also emphasize that this societal responsibility is particularly attendant when researchers and leaders select research fields and research projects and use society's resources to fund and carry out research.
3. In the daily life of a research organization, all employees must both relate to their own professional ethics and to a long list of laws and regulations that may in some way also be related to ethics. For example, many of the key ethical norms related to the involvement of humans in research have now been established in law. Here law and ethics blend seamlessly into each other. This book, therefore, addresses both professional ethics and aspects of related law and regulations that are relevant in a variety of situations encountered in daily life, without going into details that are specific to legislation in each country. The book thus provides a more complete guidance in responsible research practice (laws, rules and ethics) than the research ethics guidelines do alone. Therefore, in addition to the relevant national and international codes of conduct for research, researchers, administrators and leaders can keep the book as an aid when they have doubts about the right approach.

4. This book takes a practical perspective on research ethics. Ethics are described in connection with the tasks one has as a researcher or research leader. The focus is on lines of action that are by a clear margin within what most researchers and society in general would describe as responsible research practice. The book thus focuses on simplification and concretization, rather than problematization and theorizing (there is a rich literature focusing on the latter). Where, in an ethical dilemma, right and wrong depend on the circumstances, or the right way forward is hidden, emphasis is therefore placed on providing the reader with concrete knowledge and insight needed to find responsible avenues of action. The book also contains many examples that may be useful starting points when assessing similar cases. Some of the examples are fictional; most, however, have roots in reality but have been anonymized and rewritten to illuminate a certain ethical issue (no reference to the origins of the examples are made here). Some of the examples describe real cases, with references. All the examples originate from the Norwegian research community but are relevant for readers all over the world.

The book consists of three parts: *The first part* is a broad introduction to professional ethics in a research organization, ranging from the national and international infrastructure governing research ethics guidelines, rules and legislation, to the ethical responsibilities of organizations and individuals involved in research. This will be useful for anyone who does not have *good* knowledge and understanding of the codes of practice for research or is unsure of their own ethical responsibilities and those of their research organization. *The second part*, which is the most comprehensive, addresses issues of research ethics encountered in the workplace and in each step of a research project, from choice of topic to reporting and use of the results. *The third part* begins with a review of the extent of ethically irresponsible practice in the research community and the various forms of violation. It then describes good practices for dealing with suspicions of irresponsible conduct, from minor violations to research misconduct, and the appropriate reactions to these. The last chapters of this part are particularly relevant for whistleblowers, suspects, leaders, members of ethics committees and others who are directly involved in such cases.

The book is based on national and international guidelines for research ethics and on ethical issues that are very much at the forefront in articles and debates. It is sometimes rather detailed and often revolves around what one 'must' or 'should' do – but this is unfortunately part of the reality in research and research ethics today. This may seem like a lot to take in at once, but hopefully a comprehensive overview can also provide clarity. Since the book is not a scientific work, but a practical textbook, no broad overview of other literature is given, and references to sources are given only in connection with quotations or special information from elsewhere. However, a few references to easily accessible, complementary literature or information have been included.

I first wrote this book in Norwegian and for the Norwegian research community only. This allowed me to go into some detail concerning national guidelines, regulations and laws and other matters relevant only to Norway. After its publication in

2020 by Universitetsforlaget (Scandinavian University Press), I translated the book to English in 2021 and adapted it into an international version with worldwide relevance. However, the reader may nevertheless sense a taste of Norway in some of the examples described in this book, since they are based on my own experience and real cases I have been involved in or heard about. Despite this, I am confident that readers all over will find the examples both relevant and enlightening.

I would like to thank former professors, colleagues, collaborators and others who have contributed to the development of my own ethical attitudes and to awareness of ethical issues in research. I especially thank my colleagues at my long-time workplace SINTEF (a major, private, multidisciplinary Norwegian research and technology organization). The book is primarily the result of my participation (2007–2016) in what is now named the National Commission for the Investigation of Research Misconduct. I would therefore also like to thank my colleagues in the committee for the new knowledge and increased ethical understanding they have given me and for the many good discussions between us. A big thank you also to the staff of the Norwegian National Research Ethics Committees that effectively and committed contribute to putting research ethical issues on the agenda in the research community. Special thanks go to the late Ragnvald Kalleberg for our interesting conversations about issues of research ethics that indirectly contributed to me eventually deciding to write this book.

In connection with the *first, Norwegian edition* of the book, I would like to thank Chantal Lyche, Reidun Sirevåg, Aage Stori, Tone Fløtten, Vidar Enebakk and Anette Birkeland, who all looked over drafts of the book with a critical eye and gave me suggestions for additions, adjustments and corrections on different parts of the content. This help and support were crucial. Big thanks also to the anonymous peer reviewers appointed by the publisher, for their valuable comments and suggestions. Last but not least, I am grateful to the publisher's employees for their cooperation in the publication of the Norwegian edition of the book.

In connection with *this international edition* of the book, I would like to thank Adam King for English language editing and to Springer's team for their work in publishing the book.

Oslo, Norway   Dag Slotfeldt-Ellingsen
April 2022

Contents

Special Terminology

Research and experimental development (R&D) 'Creative and systematic work undertaken in order to increase the stock of knowledge – including knowledge of humankind, culture and society – and to devise new applications of available knowledge' (definition according to the so-called Frascati manual; see Organisation for Economic Co-operation and Development (OECD) (2015, Sect. 2.2).

Research or **science** Is used synonymously in this book as a short form of R&D, unless otherwise stated in the text (in this book it is unnecessary to distinguish between the two concepts).

Researcher A person engaged in R&D.

Research organization An organization that performs R&D, i.e. public or private universities, colleges, research institutes, hospitals, companies or organizations (or units within these, such as an R&D unit in an industrial company).

Research institution A research organization that has R&D as one of its *main tasks*, such as public or private universities, colleges, research institutes, some of the so-called research and technology organizations (RTO) and certain hospitals. Organizations that conduct R&D to a lesser extent (museums, companies that conduct R&D to support their own activities and more) are not termed research institutions in this book.

Academia Universities or colleges.

Manager (alternatively leader, collectively management) A rector, president, chancellor, principal, dean, head of institute and directors and leaders of all kinds in charge of a unit or a project within their organization.

Reference

Organisation for Economic Co-operation and Development. (2015). *Frascati manual 2015: Guidelines for collecting and reporting data on research and experimental development, The measurement of scientific, technological and innovation activities.* OECD Publishing. https://doi.org/10.1787/9789264239012-en

Part I
Professional Ethics for Researchers and Society's Expectations Towards Researchers

Chapter 1
The Ethical Basis for Research Activities

1.1 Morality and Ethics

Morality is the values, attitudes and norms that each person or community of people uses as a base for distinguishing between 'right' and 'wrong', and which influences both our own actions and our opinions about actions of others. Different people and societies can have different moralities, and morality develops over time. *Ethics* are essentially the expression of thoughts and theories of morality (moral philosophy). More comprehensive definitions can be found in easily accessible online encyclopaedias.

In modern times, morality is spoken about less and less. It is perceived as something old-fashioned and restrictive, akin to one's parents' raised forefinger. The word *ethics*, however, sounds fashionable and academic. This may explain why people increasingly choose to use the word ethics when they mean morality, and the difference between the two terms is about to vanish. In this book the two terms are also used somewhat interchangeably.

Morality is something we all acquire and develop, and it is profoundly influenced by those closest to us and by the community we grow up in and become part of. However, the origin of human morality lies probably hidden in people's basic needs and nature. Some moral norms, for example, can be seen as clever pragmatic strategies for securing our own interests within the community. Some seem rooted in human emotional life. We may choose morally good behaviour, such as fairness and honesty toward others, simply because it usually makes us feel good. However, the origins and nature of morality are a very difficult issue which have primarily been a subject for moral philosophers to ponder, since the time of Aristotle. Religions have also played a major role, not least in unifying and preserving the morality of their adherents. In the future, modern scientific methods may perhaps throw a more fact-based light on this issue, both the social aspects of morality and the chemical and biological processes involved in creating human emotions and our perceptions of right and wrong. But, as

D. Slotfeldt-Ellingsen, *Professional Ethics for Research and Development Activities*, https://doi.org/10.1007/978-3-031-25484-0_1

pointed out by the Norwegian pioneer of research ethics, Knut Erik Tranøy, the most important aspect to understand about the origin of morality is perhaps that it is not '... created by philosophers, theologians, and prophets. It is from our common everyday life that morality comes. Ethical issues are something that we all have a right to think about without having special expertise and education in ethics' (Tranøy, 1991, p. 15, translated from Norwegian by the author).

Although there is enormous human diversity, and despite the fact that the world is full of disagreements and controversies, in many moral issues within different countries and cultures there is an astonishing degree of common understanding of what is right and wrong. It is therefore possible to talk about a 'common morality' that the vast majority of people share, with some exceptions where right and wrong are disputed, and where the most principled may end up in full conflict. Internationalization has contributed to reducing the differences in common morality from country to country and culture to culture, but there are also many places in the world where tradition, religion, governance and living conditions provide the basis for divergent morality. So common morality is not completely universal, and it changes over time.

1.2   The Morality of Researchers

1.2.1   Research Ethics – A Special Kind of Professional Ethics

Researchers, like everyone else, must relate to the common perception of right and wrong in the community to which they belong. The moral views of researchers therefore reflect common morality. However, as a researcher or employee of a research organization, one will encounter some moral issues that can either be specifically related to the nature and purpose of the research or occur more frequently in a research organization. The thoughts of the research communities around this, the methodology they develop to deal with their special moral dilemmas and the ethical norms they particularly emphasize or possibly develop in order to distinguish right from wrong, constitute what can be called a special *research ethics*. This mainly concerns (The National Committee for Research Ethics in the Social Sciences and the Humanities (NESH), 2019):

1. norms that constitute good scientific practice, related to the quest for accurate, adequate and relevant knowledge (academic freedom, originality, openness, trustworthiness etc.)
2. norms that regulate the research community (integrity, accountability, impartiality, criticism etc.)
3. the relationship to people who take part in the research (respect, human dignity, confidentiality, free and informed consent etc.)
4. the relationship to the rest of society (independence, conflicts of interest, societal responsibility, dissemination of research etc.). (From the introduction on research ethics)

Research ethics can therefore be viewed as a specification of common morality within the research profession. This is not particular to the research profession – carpenters, lawyers, actors and all other professionals also have ethical issues that are distinctive or particularly relevant to their respective professions.

All professional ethics express first and foremost how the practitioners of the profession believe they should act and think to satisfy society's expectations of the work being performed, and how to proceed to ensure good relationships among the members of the same profession. Professional ethics are primarily the responsibility of the practitioners of the profession and their professional associations. However, all are aware that the authorities will intervene with laws and regulations if the professional practitioners themselves do not develop the ethical guidelines that the society finds appropriate and adequate.

Research has become a cross-border activity. This has led to an international harmonisation of research ethics, and within the international research community there is now a large degree of consensus on what is right and wrong in research.

1.2.2   Society's Expectations of Research – The Basis for Research Ethics

The starting point for almost all research today is society's need for new knowledge (in a broad sense). To provide some of this knowledge, public and private research organizations have been established. In particular, research is used as a tool in knowledge development when there is a requirement to go deep into a matter using special competence, methods and equipment, and when there are special demands on the quality of the knowledge and how this knowledge is obtained and communicated. Society has also chosen research as one of a number of instruments in university education. In turn, research organizations have employed researchers to do the job in practice, on the premises employers have established for their respective positions.

The overall societal purpose behind all research today – what society (public and private organizations and enterprises) will achieve with the money spent on research – leads to a number of general expectations being placed on research organizations and thereby on researchers. Some expectations are concrete and clearly expressed by different public bodies and actors, for example in laws and regulations that stipulate how researchers should act in different contexts, in the general conditions society lays down for research activities, and in agreements the research organizations enter into with those who finance the research. However, laypeople also have expectations about research. These are not expressed in unison, or consistently and precisely, but are nevertheless very real conditions for society's support of research. These expectations tend to be on a more general, ideal and fundamental level, such as expectations that researchers should be truthful, honest, open, objective and accurate, and from common reports about research one can gain quite a good impression of what these expectations are. In the list below, therefore, an attempt is made to put words on a spectrum of expectations that most people might join.

Society's Expectations of Research and Examples of What Researchers Will Do to Meet them
The research should be:

- Important, useful, interesting

  - Acquire knowledge and create understanding that in the short or long term is useful for society, important for the development of knowledge or enriching of individuals (this is elaborated in Sects. 6.3 and 6.4).

- Truthful, open, honest, reliable, credible

  - Describe what has been done and found (do not falsify or fabricate).
  - Describe all findings (not remove data, observations and other elements that do not 'fit')
  - Account for uncertainties, possible sources of error and alternative interpretations.
  - Be clear about who has done what (do not plagiarize).
  - Describe the research work in a way that does not mislead about the nature or importance of the work (do not distort, 'decorate' or exaggerate).

- Accurate

  - Assess the validity of all data and information derived from other sources.
  - Quality assure the research findings and especially check that all data are correct.
  - Use formulations that are precise, logical and without dual meanings – written and oral.
  - Specify the facts and assumptions that are the basis for the conclusions put forward.

- Reproducible, verifiable and traceable

  - Conduct the research in a way that reduces the risk of chance affecting the results.
  - Provide information about methods and procedures in such a way that others can control or repeat the work.

(continued)

- Provide information about what one has built on or used from previous works – be it one's own or other people's ideas, data, results, text, figures, etc.
- Keep accurate research records, save original research data and materials for a reasonable time, and make these available for any appropriate external controls.

- Factual, accountable, logical

  - Plan and execute projects so that they provide significant, relevant and reliable new knowledge or data.
  - Conclude on reasonable terms.
  - Do not go further in the conclusions than the results substantiate. Do not speculate, at least not without specifying what it is being built on.
  - Do not let personal opinions, prejudices, special interests, etc. affect the work in an inappropriate way.

- Neutral, objective, impartial, independent, unbiased, open to alternatives

  - Do not let personal political, religious, ethical, cultural, institutional or other affiliations affect the work in an inappropriate way. Let others quality assure the planning, implementation and reporting of the research.
  - Be open to the possibility that one's own theories and professional views may be wrong and those of others may be right.
  - Never bow to pressure from colleagues, managers, supervisors, clients, government agencies, etc. to carry out a research work or to adapt its conclusions in such a way that it unjustly endorses anything they wish for.
  - Provide information about funding sources and collaborative relationships (if relevant, previous relationships and support too) relevant to each research project.
  - Provide information about any recusal or conflict of interests relevant to each research project.

- Available to others

  - Give other researchers the opportunity to build on the results of one's own research.
  - Help make the results useful for many.
  - Allow others to verify the results, and to assess and control that the expectations in this list have been fulfilled.

- Considerate, respectful

  - Be aware of, take into account and take responsibility for any possible unwanted, dangerous or harmful consequences of the research.

(continued)

> - Take into account the rights, reputation and legacy of individuals.
> - Show respect for animals and nature.
> - Show respect for what scientists and others are doing now and have done in the past.
>
> • Legal, professional
>
> - Follow relevant laws, ethical guidelines and good research practices (never be negligent, never cheat).
> - Report any suspicions about irresponsible research practice to the appropriate authorities.

Meeting such expectations is largely a moral matter, and the ethics of research can in some way be seen as a 'moral contract' between the society and the research community – a promise by research organizations and researchers to do their best to meet society's expectations in a broad sense. Breaking the promise thus becomes an act of immorality.

The list above, therefore, contains keywords for the main ethical standards that the research community wishes to follow, and the listed examples of good practices are based on many of the guidelines for research ethics that the research community has established (see Sects. 2.4.2 and 2.4.3). In this way, the content of the box largely expresses the core of research ethics. Responsible research practice is essentially living up to those keywords. But there is another ingredient: a research organization's professional ethics not only include the research itself but also how each individual works collegiately and as employees in the research organization.

The remainder of this book is a detailing of how the research community and society at large believe all this should or must be done in practice, in conjunction with the many different activities and tasks carried out in a research organization.

1.3 'Responsible Research Practice' – The Standard for How Research Should Be Conducted in an Ethical Manner

Many of the methodical and ethical issues, and ethical dilemmas, a researcher may have to deal with, have previously been encountered by other researchers and have been attempted to be solved in different ways. Out from this, over many years the research community has obtained an understanding of which solutions are good and which are bad. The sum of good solutions is what is called 'responsible research practice' or 'responsible conduct of research (RCR)' (see for example Steneck (2006)). Responsible research practice is therefore the way we operationalize research ethics.

References

Steneck, N. H. (2006). Fostering integrity in research: Definitions, current knowledge, and future directions. *Science and Engineering Ethics, 12*, 53–74. https://doi.org/10.1007/pl00022268

The National Committee for Research Ethics in the Social Sciences and the Humanities. (2019). *Guidelines for Research Ethics in the Social Sciences, Humanities, Law and Theology* (English version). Oslo. The Norwegian National Research Ethics Committees. Retrieved September 21, 2021, from https://www.forskningsetikk.no/en/guidelines/social-sciences-humanities-law-and-theology/guidelines-for-research-ethics-in-the-social-sciences-humanities-law-and-theology/

Tranøy, K. E. (1991). *Medisinsk etikk i vår tid*. [Medical ethics in our time]. Bergen, Norway. Sigma forlag.

Chapter 2
National and International Measures to Promote Ethically Responsible Research

2.1 The Bureaucratization of Research Ethics in Recent Years

Initially, 'responsible research practice' was mostly *unwritten* guidelines for how one ought to or had to act in different situations. These guidelines were transferred from one generation of researchers to another – the younger ones being taught by their professors, experienced colleagues and leaders. Until the Second World War, therefore, research ethics were formalised in the form of written guidelines only to a very minor extent, but were no less present for that reason (a brief review of the history of research ethics can be found in Ruyter (2019)). Additionally, breaches of responsible research practice were mostly handled by the research community itself in a variety of ways.

Today, this is no longer sufficient. The number of researchers has increased enormously – this in itself has increased the number of deviations from good practice. Also, researchers are increasingly being recruited from cultures that may have somewhat divergent perceptions of right and wrong, and come from adolescence and educational environments where guidance and reflection on ethical issues may have had little presence. The number of international research projects is increasing and researchers now regularly move across borders. All this creates a need for clear, uniform, international ethical guidelines, and further training and supervision within research organizations.

To meet these demands, in the latter part of the twentieth century many countries began to establish committees to work out national ethical guidelines for different fields of research. An increased focus on human rights also led to national legislation related to ethical aspects of research, primarily involving living persons. In recent years a number of international organizations have also established ethical guidelines for research. In addition and in parallel, research organizations all over the world have reinforced their internal ethical training and follow-up systems and

© The Author(s), under exclusive license to Springer Nature
Switzerland AG 2023
D. Slotfeldt-Ellingsen, *Professional Ethics for Research and Development Activities*, https://doi.org/10.1007/978-3-031-25484-0_2

established ethical committees and administrative resources with research ethics assignments. All this has gradually led to the establishment of small national 'bureaucracies' that are responsible for developing and maintaining a regulatory[1] framework for research ethics, contributing to the training and awareness of research ethics and dealing with deviations from responsible practice. The majority of people involved in this are researchers and research leaders who are elected to sit on national or international decision-making or advisory committees. Some are also employed part or full-time in positions dedicated to research ethics. Research ethics has now become a profession and a specific field of activity.

However, faced with such a bureaucracy, it is essential not to lose sight of the fact that the written guidelines are primarily only clarifications of the practice that the research community has arrived at over a long period of time. In other words, the regulations are formulated by a modern research bureaucracy, but to a lesser extent created by it.

Some may find the formalization of research ethics and the many ethical committees unnecessary, signs of a lack of confidence that scientists are able to keep their own houses in order, and of a certain loss of freedom. In practice, however, it is still necessary for each researcher to make their own ethical assessments when conducting research. This is because many cases need to be assessed on the basis of the circumstances. Also, many of the issues encountered are not adequately addressed in written research ethics guidelines. New technologies and research tasks also require reflection on the suitability and adequacy of the established guidelines. Therefore, there is still a great need for each researcher to exercise ethical discretion.

2.2 National and International Actions to Promote and Secure Responsible Conduct of Research

It is a tradition that individual professionals, their employers and their professional societies, take responsibility for developing and enforcing ethical guidelines for their own profession. Within the research communities of many countries, this responsibility is attended to by *national* committees, mostly composed of experienced researchers and leaders of research organizations. These committees are usually administered and elected by national professional societies, institutions or governmental authorities, and usually supported by staff. In some countries they are organized as integral parts of a national office for research integrity, and some are even regulated by national law (examples: Denmark, Norway). The responsibilities of these committees are usually to set up ethical guidelines or codes of conduct for

---

[1] Unless explicitly stated otherwise, in this book 'regulations' is used as a collective term for all kinds of laws, regulations, rules, guidelines, codes of good practice, etc. that researchers ought to or are obliged to follow in their work.

research in general or research in specific fields. In some cases, these committees and their supporting staff may have additional responsibilities, such as advising the national research community concerning ethical issues, promoting good research practice through various measures, coordinating their work internationally, etc.

Modern research is carried out across borders to a large extent, and the reporting and use of research results are largely international. This makes international coordination necessary, and has led to a number of *international* initiatives. Some examples:

- To coordinate the work of the national research ethical offices in Europe, the European Network for Research Integrity Officers (ENRIO) was established in 2007. ENRIO facilitates information exchange and experience transfer between countries.
- In Europe, major international organizations such as the European Commission, the OECD, the European Science Foundation (ESF), the All European Academies (ALLEA), and others have also established European ethical guidelines that largely address the same issues of concern as in each country, albeit based on their own interests and needs. These organizations employ international committees in their work, thus becoming important meeting places for personnel involved or interested in research ethics.
- Globally, the World Conferences on Research Integrity have played an important role in the exchange of experience on research ethical issues between countries. From the first one in Lisbon in 2007, these conferences have taken place every second year, and have, among other things, helped develop international guidelines for research ethical issues, such as the Singapore Statement and the Montreal Statement (see Sect. 2.4.3).

The small national and international bureaucracy dedicated to research ethics, on the side lines of the research itself, has been established to raise the awareness among researchers and research leaders that ethics and professional quality go hand in hand, and to help them act and make choices in a way that can be ethically justifiable. On the other hand, one may ask whether this well-intentioned bureaucracy renders researchers and research leaders passive. It is easy to think that 'the job is done' when the relevant ethical guidelines are readily available online, when the institutions have followed up and ensured that all students and new employees have attended courses in research ethics, etc. But it isn't. The research bureaucracy and its committees, staff and guidelines are only there to help. The responsibility for acting ethically in research lies entirely on the individual researcher and research leader, and ethical reasoning is part of their daily activities, integrated into the research and management tasks. This is discussed in more detail in Chap. 3.

2.3   Research Ethics as a Growing Research Field

In parallel to the formalization and bureaucratization of research ethics, research ethics is growing as a separate research area, and a number of researchers in a variety of fields are targeting this area. Research-funding bodies are also increasingly interested in supporting research and innovation related to research ethics and ethical aspects of the interaction between research and society. Several journals devoted to research ethics have also been launched (a majority in medical ethics and bioethics). The same applies to conferences and meetings specifically aimed at research ethics. At this early stage it is relevant, however, to form an opinion on what new knowledge and increased understanding the new research field may generate, what priority it should be given and to what extent its growth should be encouraged. Not everything new is important or useful.

Increased knowledge on ethical issues can help the research community identify ways and means that promote good research practices. At the same time, the theorizing and problematization that research often centres around may in some cases complicate rather than clarify what is right and wrong in specific issues one encounters as a researcher.

2.4   The Regulatory Framework for Conducting Responsible Research

2.4.1   The Regulations Everyone Must Follow in Research

People often act on the basis of their own ethical views, but are judged on the basis of others'. This can lead to situations where one believes oneself to have acted correctly while most others see the act as immoral and blameworthy. The condemnation from the research community may then be harsh, and one's future as researcher may be at stake. This is where the national and international guidelines for research ethics come in. By having such written guidelines, all researchers can know in advance what the research community considers to be right and wrong, and can, by making themself familiar with the guidelines, avoid acting reprehensibly due to ignorance or thoughtlessness. As will be discussed in further detail later, the ethical guidelines for research not only apply to researchers but also to all personnel involved in research and to organizations where research is carried out.

In addition to the national and international guidelines for research ethics, all research organizations will have internal procedures, perhaps even supplementary ethical guidelines, to ensure that their own employees act in accordance with the expectations of the society and the research community. On top of this come national laws and regulations that clearly state societal expectations on particular issues. The regulations that everyone involved in R&D has to deal with, are therefore threefold; see the box below. Each class of regulation is discussed in greater detail in what follows.

The Regulations that Everyone Involved in R&D Must Follow

- National and international guidelines for research ethics.
- Internal procedures and guidelines in the research organizations.
- National laws and regulations.

It is a basic principle in the research ethics that individual researchers have a personal, independent responsibility to familiarize themselves with all this. If something goes wrong, it will therefore be of little help to blame poor or inadequate training at educational institutions or the workplace. It is therefore wise to choose procedures that are well within what the research community considers good practice. The research community can react strongly to mistakes.

2.4.2   National Guidelines for Research Ethics

National guidelines for research ethics, or codes of conduct for research, differ in setup, scope and formulations from country to country, but the underlying ethical norms and the practical advice given are largely the same. However, in a book like this – written for researchers in many countries – these differences make it impossible to treat national ethical guidelines in detail. Instead, the readers must, on their own initiative, familiarize themselves with the guidelines of their respective countries.

All guidelines for research ethics are periodically updated and occasionally complemented with guidelines for new or special research fields. To stay up to date, a good habit may be to visit the websites of relevant national and international codes of conduct for research, for example once a year.

2.4.3   International Guidelines for Research Ethics

Many international organizations have seen it appropriate to develop guidelines for research ethics related to their own needs. The purpose is partly to ensure that the research activities they are involved in are carried out responsibly, and partly to establish a research ethics standard across national borders. Some of these international guidelines apply to special fields of research (primarily in medicine), while others apply to all disciplines. Like the national guidelines, the international ones are regularly updated and new ones appear. Each researcher should familiarize themselves with the guidelines most relevant to their own activities. Below are some examples of international guidelines that apply to *all* fields of research:

- The European Code of Conduct for Research Integrity (ALLEA, 2017). This is a set of guidelines for research ethics prepared by ALLEA, originally from 2011, but significantly revised in 2017. These guidelines focus on issues that the research community is particularly concerned about. They are not intended as a substitute for national guidelines, but express an agreement on principles and priorities of research ethics in the European research community. Projects supported by the European Commission are to be based on these guidelines. As an expression of responsible research practice in Europe, the code should be looked upon as normative for research in Europe, and as a possible guide for researchers in other countries as well.
- The European Charter for Researchers (European Commission, 2005). This was a recommendation from the European Commission to the European Union member states, adopted and published in May 2005 together with the Code of Conduct for the Recruitment of Researchers. The Charter has been adopted by a number of research organizations in Europe. It places demands on researchers, research institutions and funding organizations. In particular, the Charter's perspective views research in terms of the interests of the society and on the relationship between the researcher, employer and source of funding. It should be looked upon as normative for research in Europe, and as a possible guide for researchers in other countries as well.
- Singapore Statement of Research Integrity; hereafter called the Singapore Statement (World Conferences on Research Integrity, 2010). This statement lists a number of principles of responsible research practice. They were worked out in conjunction with the 2nd World Conference on Research Integrity in Singapore and published in September 2010. This was the first international effort to develop uniform guidelines and rules of good practice in all kinds of research worldwide. The statement is not binding for anyone, but is intended as a starting point for further work on creating principles and guidelines that can be used regardless of subject and type of research. Although the statement is non-binding, it should be looked upon as a description of good, international, research practice.
- Montreal Statement on Research Integrity in Cross-Boundary Research Collaborations; hereafter called the Montreal Statement (World Conferences on Research Integrity, 2013). This statement lists a number of principles of good research practice that were worked out in conjunction with the 3rd World Conference on Research Integrity in Montreal in 2013. The statement covers general guidelines for research collaborations that cross national, institutional, disciplinary and sector boundaries. Although the statement is non-binding, it should be looked upon as a description of good, international, research practice.
- The International Committee of Medical Journal Editors' (ICMJE) recommendations for *co-authorship*, usually called the 'Vancouver recommendations' (ICMJE, 2019, December, II. A. 2., Who is an author?). This is a set of criteria for being a co-author on a scientific article which was initially devised for medical research but is now widely used in other disciplines as well. It forms part

of Recommendations for the Conduct, Reporting, Editing, and Publication of Scholarly Work in Medical Journals, prepared by the International Committee of Medical Journal Editors. These recommendations are based on the initiative of an informal group of publishers who came together in Vancouver in 1978, hence it being known as the Vancouver recommendations. They are updated regularly. They are described in more detail in Sect. 16.6.1

International guidelines for research ethics are generally less detailed and comprehensive than national ones, and each has their own content, scope and formulations. Ignoring these differences, both international and national guidelines express the same overall, basic principles for responsible execution of research. Nevertheless, the differences may cause some confusion. On the other hand, reading the homeland guidelines, as well as other national and international guidelines, can often give a broader understanding of the different ethical issues encountered.

2.4.4 Guidelines for Research Ethics – A Combination of Guidance and Requirements

There is a big difference between guidelines for research ethics and legislation. The former are primarily aids in exercising responsible research practices, whereas the latter are rather precise specifications for what to do and what not to do. NESH (2019) has expressed it as follows:

> The guidelines for research ethics do not serve the same role or function as legislation. The guidelines primarily serve as tools for researchers and the research community. They identify relevant factors that researchers should take into account, while acknowledging that researchers often have to weigh such factors against each other, as well as against other requirements and obligations. (From the introduction)

This expresses that in special contexts or circumstances it may be right, and thus ethically acceptable, to deviate somewhat from the guidelines. A prerequisite must be to be open about what one does differently, and be able to justify it well. It is also important to understand that research ethics guidelines are not as precisely formulated as laws. If one is unsure of how to interpret a particular guideline, it is therefore often better to look for the intentions behind the guideline than to embark on detailed interpretations of words and formulations.

Despite this, many will probably perceive the guidelines for research ethics as rules one *must* follow. For example, no one will accept violations of the basic, research ethics norm of truthfulness, because it is in a way central to the 'definition' of research. When suspicions of serious deviations from responsible research practice are investigated by institutions or commissions, the violations under investigation will initially always be assessed in relation to what is explicitly stated in the relevant national guidelines. Therefore, it is wise to look at the ethical guidelines as requirements for the research profession, but at the same time have in mind that they do not cover everything, that in special contexts there may be good and ethically justifiable reasons to deviate from them, and that right and wrong sometimes depends on the circumstances. In other words, in many situations, researchers must also use their own ethical discretion, justify the assessments they make and be responsible for the decisions they end up with. An example of the importance of specifying and justifying any deviations from guidelines for research ethics is given in the box below.

Guidelines for Research Ethics Are Expected to Be Followed in All Activities Referred to as R&D: An Example

In a case of plagiarism at a private research institute, a national investigating commission found that in several commissioned R&D reports, text and figures from previous institute reports (partly written by the same authors) were used verbatim without references. This violates important expectations for research, such as truthfulness, traceability, openness and respect for the contributions of other researchers. In a scientific journal article, this would have been considered plagiarism, but the research institute where this took place explained and reasoned that their commissioned R&D reports were of a special character, and that repetitions of text and figures from previous reports, without references, were in this case acceptable practice. However, the reader was not in any way informed of the deviation from the standard of good quotation and referral practices in research, despite the fact that the reports appeared to be regular research reports, were part-funded by public research funds and written by researchers at a research institute. The commission therefore found that the institute's view was unacceptable – readers could be misled. They expressed themselves this way (The Norwegian National Commission for the Investigation of Research Misconduct, 2012, online edition, p. 35, translated from Norwegian by the author): 'When the institute "hoists the research flag", it is common good scientific practice [...] that must be followed.' If the institute had clearly stated and reasoned that the reports deviated from good citation and referral practices, the conclusion on this would probably have been different.

In newspaper articles and debates on research and research policy, research ethics norms are sometimes used as an argument for different points of view. Some then seem to raise the guidelines for research ethics, or some of them, to binding rules of action. Taking a dogmatic approach to ethical norms usually goes awry. There are few general answers to what is ethically right and wrong in every situation. All cases must be considered separately, and different research ethics norms and non-ethical factors must be weighed against each other in order to find the best position or alternative of action that can also be defended ethically. Dogmatism in research ethics, which can to some extent be compared to religious, ideological and political dogmatism, is unscientific in its essence because it is uncritical and thus only to a limited extent truth-seeking and open to divergent views.

2.4.5   The Relevance of the Guidelines in Different Types of Work

The guidelines of research ethics discussed above are basically designed to provide guidance on all kinds of private and public research and research-related work. The guidelines are also relevant regardless of how the research is funded, but research carried out on commission for others raises special ethical issues that are discussed in more detail in Chap. 9.

Most research organizations are also engaged in activities other than research and development – for instance, teaching, dissemination of research, work in connection with putting research results to use, etc. Many also conduct surveys, engineering, laboratory services, etc. on commission for clients. These tasks are usually closely related to the research work itself, and the general guidelines of research ethics are therefore basically well suited here as well. However, such activities can raise particular ethical issues. Some of them are discussed in Sect. 5.13 (supervision of students), Chap. 9 (R&D on commission for clients), Chap. 18 (commercialization of research results) and Chap. 20 (dissemination, participation in public debate).

From time to time, some researchers are also engaged as consultants, advisers, experts or similar. Here it is not as obvious that the work should be based on guidelines for research ethics, and little is said about such activities in national and international research ethics guidelines. Two examples of consulting work carried out by researchers may give an indication of what the answer should be:

Example 1: A small manufacturing industry asks a research institute to calculate the loads on one of the company's mechanical products – a typical technical consultancy job. The report is to be used as documentation for the product's properties. They expect a report that is truthful and accurate; where strengths and weaknesses of the construction are clearly indicated; where the work is quality assured; where the uncertainties are discussed; where the calculations are described in such detail that they can be controlled or repeated by others; where the researchers are open about their neutrality, etc.

Example 2: A law firm engages a law professor to prepare a report about how a new EU regulation can affect the business of a group of the firm's clients. They expect a report that is truthful, factual, justifiable, neutral; where difficulties and opportunities arising from the new regulation are clearly stated; where the professor is open about what assumptions she makes for the assessments, etc.

The examples suggest that the ethical norms for research are initially both relevant and well suited for certain types of consultancy assignments. But not for all. For example, an advertising consultant must also deal in the truth about a product, but on the basis of ethical guidelines for the profession that will differ from the guidelines for research. In other words, if researchers are also to be guided by the research ethics in consultancy assignments, there will be certain assignments to which they must say *no*.

In conclusion, research ethics seems well suited as an ethical basis for a wide range of activities that researchers have particularly good competence, equipment, etc. to get involved in. The certainty that researchers follow research ethics guidelines when performing works on commission for others is probably also a compelling reason why clients approach research communities with their projects – be it with research or with other tasks. Because the market for consultancy and advisory services is dominated by a large number of companies and individual enterprises with their own ethical guidelines, it is also important that researchers and research organizations that undertake such assignments explicitly state that they always follow the guidelines for research ethics.

However, the main reason why researchers should follow the same ethical guidelines in everything they do – from research to consultancy – is that this is *necessary to protect their credibility as a researcher.* Operating with different ethical norms according to the character of the activity will confuse and stir doubts about the ethical basis for their main activity as a researcher.

2.4.6   The Relevance of the Guidelines for Different Job Categories

Researchers
With formulations such as 'The researcher shall ...', both national and international guidelines for research ethics primarily address the *individual researcher.*

One might then ask: What is a 'researcher'? The title can obviously not include everyone who has received a formal research education. A director general of a ministry, a CEO of an industrial company or a taxi driver, all with a PhD, are not expected to work according to the standards of research ethics. For them, other professional ethics apply. However, in this context the term researcher should include everyone who is *employed in a research position or a position where research is part of the job* (in research organizations one will find employees with different formal education in such positions – 'researcher' is not a protected title). Likewise, it is obvious that students working on a research thesis (for example as part of a master's or PhD degree), must follow the guidelines for research ethics that are relevant for their research field. It is also reasonable to require that anyone who is not in a research position, but who works with something that methodically and otherwise can be called research, and who presents their work as research, must follow the guidelines for research ethics (unless they explicitly state otherwise).

Others Who Participate in Research Work or in Different Ways Are Part of the Research System
In order to rely on the results of a research project, it is important that not only the researchers, but everyone who is involved in it in different ways – be they managers, secretaries, engineers, technicians, students, etc. – also follow the research ethics guidelines, as far as relevant.

The national and international guidelines for research ethics are not as clear about this as they are on the ethical responsibilities of researchers and research institutions, but common sense tells us that it must be so: For example, everyone will find it unacceptable that a laboratory technician fabricates some extra points in a series of measurements in order to finish quickly, that a research administrator misleads a source of funding about the importance of a research project, or that a student who is not formally employed as a researcher, but who participates in a research project, tries to impress with good arguments and formulations that are essentially plagiarized from others. Truthfulness, transparency, accountability, accuracy, etc. must reasonably be expected by everyone in and around a research team.

As a consequence, research organizations must provide relevant education and tutoring in research ethics to *all employees* who are somehow involved in the research activity. They must also react and implement sanctions in the event of deviations from good research practices when it is done by others than the researchers. Because little is said about 'non-researchers' in the national and international guidelines for research ethics, it is probably both necessary and appropriate that each research organization establishes internal guidelines for this. However, here too there may be some job-specific ethical issues that need to be dealt with specifically.

Organizations that do not conduct research themselves, but have important tasks within what one might call the 'research system' (a research council, scientific publishing houses and others), may also be directly involved in the research activities, for example in reviewing projects and scientific works. Also in these circumstances the staff should be trained in research ethics and follow the guidelines for research ethics, whenever relevant.

2.4.7   Research Organizations' Own Procedures and Ethical Guidelines

Research organizations, research councils, sponsors and clients, publishing houses, project teams, etc. may find it appropriate to establish their own more detailed codes of conduct or ethical guidelines for special issues. There is also a clear trend of research organizations and others in the research system establishing a greater and greater number of *internal procedures* for how their own employees *have to* act in different situations.

The rationale is partly that national guidelines for research ethics are not exhaustive enough or do not take into account special circumstances on an issue, and partly that procedures are better than guidelines when it comes to ensuring that employees do not make mistakes or move into the grey zones towards unacceptable conduct. Internal procedures are rooted in legislation, research ethics, internal corporate governance and internal organizational and administrative systems. On the one hand, internal procedures can make daily work easier for employees (they can often just follow a template for good case management). On the other hand, an unfortunate side effect can be that the ability to exercise independent, ethical judgment is weakened.

2.4.8   National Legislation

Until the middle of the twentieth century, the authorities considered that research communities' own assessments of right and wrong, and the self-regulation they exercised, were sufficient to ensure that the research was carried out in the best interests of society, without unacceptable harm for anyone or anything. However, certain incidents, particularly in medical research, led to a gradual introduction of a number of laws that directly regulate how research in some subject areas should be carried out, and in some areas limits what kind of research can be conducted. The new legislation related to research, which evolved from the latter half of the twentieth century, primarily concerns research involving humans and animals, and research that can lead to the manipulation of nature and organisms (including humans). This was triggered by information that was revealed to the general public about experiments that had led to great harm for the research subjects carried out during and after the Second World War. These experiments were often carried out on prisoners, soldiers, the poor, psychiatric patients and other weak and unprotected individuals, often without their consent, without them being informed about the risk of being harmed, and often with the endorsement of the authorities. Changes in views on human rights led to societal intervention by means of legislation.

In general, it is not difficult to imagine that the desire or need to acquire new knowledge can motivate scientists so strongly that they push the limits of what are acceptable consequences and acceptable risk. This suggests that these limits should, in some areas, be set by society, in terms of laws and regulations, rather than by the research community itself. Increased legislation related to research is consistent with a trend in which society increasingly resorts to the law to ensure that different actors in society operate according to common moral norms. Society is likely to view research in the same way, and in the years ahead various aspects of responsible research practice may find their way into laws and institutional procedures.

Some researchers are against more and more elements in the guidelines for research ethics being included in legislation. They are of the opinion that the research community itself should and can keep order in its own house. Yet, decades after written guidelines for research ethics were introduced in most countries, the number of deviations from the guidelines is still too great (see Chap. 21). The inability of the research communities to implement the guidelines better can be seen as a justification for the authorities to intervene, at least in areas of great importance to society.

Violations of laws and regulations are of course also deviations from responsible research practice. The laws relevant to research vary from country to country, although international cooperation and trade have contributed to harmonisation between the nations. Since law and ethics often overlap, the legal differences from country to country sometimes lead to differences in the respective national guidelines for research ethics. All researchers therefore have an independent responsibility to clarify the legal provisions that may be relevant for their own activities.

The box below shows examples of areas and issues where national laws contain provisions with which researchers and research organizations must comply.

Examples of Areas of Legislation Relevant to Researchers and Research Organizations

- Laws governing how research in certain areas should be carried out or restricted, including fields and issues such as:

  - Research involving humans, i.e. issues related to human rights, privacy, informed consent, confidentiality, preapproval of projects, processing of personal data, human remains, etc.
  - Aspects of medical/biomedical research and gene technology.
  - Research involving animals, cultural heritage, nature and nature's diversity, etc.

- Laws relevant to organizational, administrative, operational and personnel matters at a research organization, including issues such as:

  - Employee rights, academic freedom, working environment, harassment and bullying, corruption, whistle-blowers, administrative procedures in public and private organizations respectively, public insight and access to documents, transparency, etc.

- Laws relevant to copyright, inventions, etc., including issues such as:

  - Copyright of original scientific works, intellectual property, the right to inventions originating from employees, patenting and patents, etc.

- Laws of special relevance for commissioned research, including issues such as:

  - Contracts, tendering, etc.

2.5 When Laws, Regulations and Ethical Guidelines Do Not Provide Unambiguous Answers: Discretion in Research Ethics

Neither laws and regulations nor guidelines for research ethics provide answers, let alone unambiguous answers, to every ethical issue one might face in R&D activities; see Fig. 2.1. Where there are voids, obscurities and alternatives, assessments of what is right and wrong must *be based on the exercise of discretion*. In everyday life, *discretionary decisions are made all the time*.

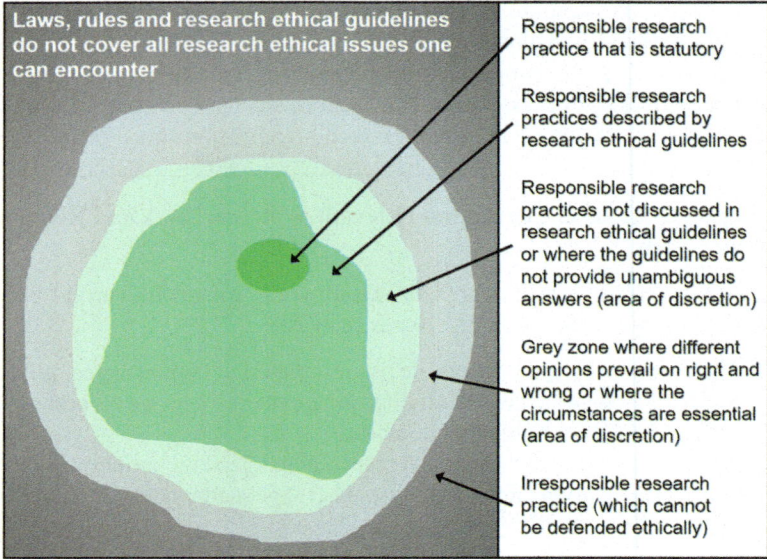

Fig. 2.1 When responsible research practice must be based on ethical discretion

The framework for exercising discretion in *research ethics* is less well established than the framework for exercising *judicial* discretion, where there are relatively well-established guidelines and precedents for this.

2.6 The Further Developments of the Regulations – Will the Best Become the Enemy of the Good?

There is a clear trend for increasingly comprehensive, detailed and specified research ethics guidelines to be developed, for ever more research ethical issues to be covered by legislation, and for these laws to become ever more detailed. The trend is reinforced by the fact that it is occurring internationally.

It is important to understand how this happens. Often there are one-off episodes and coincidences that trigger new measures. An example: A few years ago, when it was discovered that a Norwegian researcher had fabricated data on a large scale, the Norwegian government quickly established a new, permanent, national committee to investigate research misconduct. At the same time the concept 'research misconduct' was defined by law in Norway. The appropriate minister took action – for when individuals do something wrong, it is common to evaluate the entire 'system'. One asks whether the training, guidance, procedures, quality system, control and management are good enough. And everything can be improved. In such cases, a 'requirement' often builds up both within the organization and from the outside, that responsible leaders must take action – and so the leaders do precisely that. The solutions they then often resort to are more guidance or governance, and the regulations become swollen. The danger is that over time this can lead to a patchwork of guidelines and rules both nationally and within each research organization, which can be unwieldy to navigate. Some are also concerned that regulations in certain areas may become too detailed and inflexible in relation to the diversity of research tasks – and more bureaucratic than ethical. One is increasingly led along a trail of tasks, and when all have been ticked off, one has passed the finish line and acted ethically. Along the trail, the focus is to carry out a task according to a given procedure and move on to the next task. The associated ethical issues are easily forgotten. This can have a weakening effect on the individual's ethical judgment – where no trail has been made, one is left bewildered. The purpose behind such gradual changes is always good, and taken in isolation the measures can be highly necessary. But looking at the totality and reflecting on the development over time, one can ask whether the best can eventually become the enemy of the good. For lawmakers, members of ethics committees, leaders of research organizations and others who are jointly responsible for managing the entirety of the regulations, simplification and coordination may therefore probably become more and more important.

References

All European Academies. (2017). *The European code of conduct for research integrity* (revised ed). ALLEA. Retrieved September 21, 2021, from https://allea.org/code-of-conduct/

European Commission. (2005). *The European charter for researchers.* Brussels. European Commission. Retrieved September 21, 2021, from https://euraxess.ec.europa.eu/sites/default/files/am509774cee_en_e4.pdf

International Committee of Medical Journal Editors. (2019, December). *Recommendations for the conduct, reporting, editing, and publication of scholarly work in medical journals, section II.A.2 Who is an author?* Retrieved September 21, 2021, from http://www.icmje.org/icmje-recommendations.pdf

The National Commission for the Investigation of Research Misconduct. (2012). *Granskingsutvalgets uttalelse i sak om vitenskapelig uredelighet ved et teknisk-industrielt forskningsinstitutt.* [The Investigation Committee's statement on scientific misconduct at a technical-industrial research institute]. Oslo. The Norwegian National Research Ethics Committees. Retrieved September 21, 2021, from https://www.forskningsetikk.no/contentassets/9c30a4e0b4714729984273f88f520930/granskingsutvalgets-uttalelse_nettutgave.pdf

The National Committee for Research Ethics in the Social Sciences and the Humanities. (2019). *Guidelines for research ethics in the social sciences, humanities, law and theology* (English version). Oslo. The Norwegian National Research Ethics Committees. Retrieved September 21, 2021, from https://www.forskningsetikk.no/en/guidelines/social-sciences-humanities-law-and-theology/guidelines-for-research-ethics-in-the-social-sciences-humanities-law-and-theology/

Ruyter, K. W. (2019). *The history of research ethics* (English version). The Norwegian National Research Ethics Committees. Retrieved September 21, 2021, from https://www.forskningsetikk.no/en/resources/the-research-ethics-library/systhematic-and-historical-perspectives/the-history-of-research-ethics/

World Conferences on Research Integrity. (2010). *The Singapore statement of research integrity.* Retrieved September 21, 2021, from https://wcrif.org/statement

World Conferences on Research Integrity. (2013). *The Montreal statement on research integrity in cross-boundary research collaborations.* Retrieved September 21, 2021, from https://wcrif.org/guidance/montreal-statement

Chapter 3
The Allocation of Ethical Responsibility for Research

3.1 General

The responsibility for ensuring that a research work is planned, carried out, reported and followed up in accordance with the regulations discussed in Chap. 2 rests on everyone who is in some way involved in the work: Researchers and others who carry out the work, project managers and line managers, the supporting administration, any partners, clients, sources of funding, publishing houses, etc. The responsibility is not the same for everyone. This chapter focuses on the ethical responsibility that lies with the individual researchers and leaders in the research organizations.

3.2 The Independent Ethical Responsibility of the Researcher

The individual researchers' personal responsibility for the research they conduct has its roots far back in time when research was almost always carried out on an individual basis. Today, the situation is different. Now, research takes place within research organizations, often based on the needs and interests of funding sources and clients. Today, the research is also very often carried out in cooperation with researchers at one or more institutions and often with the participation of students, public or private stakeholders and others. However, none of this unsettles the principle that the individual researchers have an independent, ethical responsibility for their own choices and actions in a research context.

The ethical leeway within which researchers can act is defined by established practice and custom in the research community, as well as written, national or international guidelines for research ethics. This includes relevant laws and regulations. In addition, the individual researcher may also have other obligations of an ethical

© The Author(s), under exclusive license to Springer Nature Switzerland AG 2023
D. Slotfeldt-Ellingsen, *Professional Ethics for Research and Development Activities*, https://doi.org/10.1007/978-3-031-25484-0_3

nature through employment contracts, agreements with clients, funding sources, publishers, etc.

3.3 The Researcher's Responsibility to Internal and External Partners

When researchers work together, questions related to the sharing of responsibilities and the limits of the individual researcher's responsibility arise. Some of the issues often encountered in this respect are discussed in the sections below.

3.3.1 Responsibility for the Whole in a Collaborative Project

In most research collaborations, each participant has well-defined tasks that the parties agree on before the collaboration begins. The tasks each partner undertakes become their responsibility. For the research community and all other users of the research results, the individual tasks are often of less interest – it is first and foremost the whole that matters. Therefore, someone must always be responsible for the whole in all phases of the project, from planning to reporting and the possible use of the results.

In collaborative research, the common practice is for the partners in the collaboration to take collective responsibility for the whole. This is clearly stated in the aforementioned Montreal Statement (World Conferences on Research Integrity, 2013):

> Collaborating partners should take collective responsibility for the trustworthiness of the overall collaborative research and individual responsibility for the trustworthiness of their own contributions. (§ 1 on Integrity)

The Montreal Statement applies to cooperation that crosses the boundaries between countries, institutions, disciplines, etc., but the principle that each researcher has a dual responsibility – for their own work and for the whole – must also reasonably be regarded as a normal allocation of responsibility in cooperation between researchers within one project group, research unit or research organization.

The question, then, is what this collective responsibility entails in practice. The answer must be given on the basis of discretionary assessments in each project, but certain tasks stand out as natural elements of a collective responsibility:

• Responsibility for ensuring that the project plan attends to all relevant scientifical and research ethical issues in the project (see Sect. 8.2) and describes how the work will be carried out, quality-assured and controlled (scientifically, financially and administratively) in an ethically responsible manner. Likewise, responsibility for ensuring that the cooperation agreement between the parties clearly expresses the rights and obligations of each participant (see Sect. 8.3.4). These two documents must be agreed upon before the project can start, and all participants will therefore have insight into them and joint responsibility for their content.

- Responsibility for ensuring that project reports, journal articles, conference contributions, etc. are presented and published in accordance with responsible research practice (see Sect. 16.9). The reporting and presentation of the project results are matters that the collaborating partners must agree on, and for which they therefore have a joint responsibility – regardless of who has carried out the work described and written the text.

- Responsibility for ensuring that any other joint tasks that may have been agreed upon are carried out in a responsible manner.

These examples are related to tasks within a collaborative project that involve more or less all participants, and which therefore naturally form the core of their collective, scientific and ethical responsibility. It is important to note that with such an understanding of the collective responsibility, each project participant will be held liable to a limited extent for professional errors or violations of research ethics norms committed by other project participants in their assigned work tasks. This is a natural consequence when each participant works independently with their subtasks, which is common in research. Research collaboration is based on mutual trust between the participants. Unless otherwise agreed, it is therefore common for each participant to be responsible for ensuring the quality of their own work. If so, the participants will have very limited insight into the details of each other's work. If it is later revealed that one of the participants has, for example, falsified data, it will normally be unfair to make the other partners co-responsible for the misconduct, as long as they have taken care of their collective responsibilities, as described above. This view has often (but not always) been asserted by investigative committees in a number of cases of research misconduct. In these cases, it was found that the person who broke with good research practice also misled the partners. However, in some of these cases, the partners, or some of them, were criticized for lacking due diligence and for their passivity in the overall quality assurance of the project. More specifically, co-authors have in some cases been blamed for what other co-authors have done wrong. To be fair, such conclusions must be based on individual assessments of each partner and co-author, both in terms of the objective elements of the case and in the question of the degree of guilt (see Sect. 23.2) – having a collective responsibility in a collaboration does not imply that one is automatically guilty of irresponsible actions carried out by others in the collaboration.

Although such task-split and *trust-based* collaborations have worked well for many years, questions can be raised about whether they will work well in the future. Quality control of one's own work can provide acceptable quality assurance when carried out by competent and honest researchers, but not when carried out by professionally weak or fraudulent researchers. At a time when breaches of responsible research practice no longer seem to be uncommon (see Chap. 21), research teams should consider doing more to safeguard the scientific and ethical quality of their projects (quality assurance and control are discussed in more detail in Sect. 4.2.2).

3.3.2 Responsibility as Project Manager, Lead Author, Etc.

In most collaborative projects, some participants will mainly work on sub-tasks while others have more to do with the whole. The initiators or planners, the project manager and any lead author are examples of the latter. With tasks comes responsibility, but one can ask whether, for example, a project manager or a lead author has a greater research ethical responsibility for the whole than the other collaborators. Although neither national nor international guidelines for research ethics provide an unambiguous answer to this, it is reasonable that all partners and co-authors share responsibility for the *whole* equally (as described above), *unless they agree otherwise*. On co-authors, ALLEA (2017, Sect. 2.7) lends support to this in stating: 'All authors are fully responsible for the content of a publication, unless otherwise specified'.

Such an agreement may, for example, give the project manager, a lead author or another person or group a specified responsibility for verification or quality assurance and control of the project work or parts of it, and be afforded working conditions that make this possible. The other participants in the project will then naturally have correspondingly less responsibility if it is found that one of the participants has acted irresponsibly, provided they have exercised due diligence and possibly reported suspicions of inappropriate behaviour to those responsible for the project.

3.4 Ethical Responsibility of Research Organizations and Research Leaders

Like all other organizations, the research organizations have an overall responsibility for all matters in and related to their own activities. The responsibility lies with the leaders of the organization. Both national and international guidelines for research ethics specify elements of this responsibility and indicate that in practice researchers are responsible for their own research, while research organizations are particularly responsible for how the research is facilitated (the overall prioritization, staffing, organization, administration, systems for quality assurance and control, overall management and control, etc.).

When it comes to ethics, one of the main responsibilities of the research organization is to ensure that all employees have sufficient knowledge of and good attitudes toward research ethics, and to ensure that anyone who directly or indirectly participates in the research is familiar with the relevant research ethics guidelines. It should be emphasized again here that this not only applies to researchers but to everyone who in one way or another is involved in the research.

In addition to this, a number of organizational, administrative and research-specific policies and procedures must be in place to ensure that the organization operates in an ethically sound way. Some of these are established in laws and regulations, some are described in national and international guidelines for research ethics, some originate from general organizational practices and some will simply be based on common sense. The box below summarizes the most important measures. They will also be described in further detail later in this book.

Important Elements of the Research Ethical Responsibility of a Research Organization

- A research policy that is in accordance with the overall societal responsibility of the organization (this should also include related strategies for teaching, dissemination, etc.). In this context it is important to have:

  - An overall research plan and a strategy for the development and use of the organization's resources that are ethically justifiable from a societal and scientific perspective.
  - Procedures for following up on plans and strategies, and leaders steering in the right direction.

- An overall operation of the organization that is in accordance with responsible research practices. Important in this context is to have:

  - Leaders and administrative personnel who are well trained in and reflective on research ethics.
  - Leadership focus on responsible research practice.
  - Management and administration that lead by example and allow truthfulness, accountability, transparency, etc. to characterize their work.

- An administrative system that facilitates responsible research practice and prevents anyone from acting irresponsibly. In this context it is important to have:

  - Clear requirements for employees' ethical competence and arrangements for training employees in research ethics when necessary. This applies not only to researchers, but to everyone involved in research.
  - Procedures to ensure that supervisors (for students, new employees, etc.) are well qualified in research ethics and understand that mentoring in research ethics is an integral part of the job.
  - Procedures to ensure that students, guest researchers and others who participate in the organization's research activities have sufficient competence in research ethics.
  - Procedures, guidelines and personnel support related to research ethics, adapted to the organization's activities and resources.
  - Procedures for conflict management that are able to prevent collegial disagreements from destroying a working environment and triggering ethically unacceptable behaviour.
  - An organizational culture and working environment that promote responsible research practice.

(continued)

- A system to ensure that research ethical issues are addressed in agreements with clients and funding sources. In this context it is important to have:

  – Procedures and contract templates for agreements with clients and funding sources that, in addition to all other relevant matters, also address the ethical issues in the project and in the relationship between the contracting parties.

- A system for checking that responsible research practice is followed. In this context it is important to have:

  – Procedures for quality assurance and control that not only ensure the quality of the research work, but also contribute to preventing and revealing deviations from responsible research practice. Quality assurance procedures that include more than the researchers' own control and the journals' quality control may then be necessary.
  – Procedures for project organization and management where the project participants' research ethical responsibilities are clearly defined.

- A policy for responding to all suspicions of breaches of responsible research practice. In this context it is important to have:

  – Procedures for reporting ('whistleblowing') suspicions of breaches of responsible research practice and other matters worthy of criticism, and procedures for handling such reports.
  – Procedures for handling various deviations from responsible research practice (from the smallest to the most serious).
  – Procedures for notifying those affected by wrongdoings, for damage-limiting measures and for transparency in the case of serious deviations from responsible research practice.
  – Procedures for sanctions against persons who have acted improperly.
  – Procedures for system auditing in the aftermath of serious deviations from responsible research practice.

3.5 Organization of Work Related to Research Ethics Within a Research Organization

All research organizations are split into operational units, often on several levels (faculties, institutes, divisions, departments, sections, laboratories, libraries, etc.). The managers of each unit are responsible for the activities within their unit, while the top manager and the board have the overall responsibility for the whole. Each manager thus has an independent responsibility to ensure that the unit follows responsible research practice.

All organizations have staff and support functions to help and assist their operational units. In the context of ethics, this will typically include:

- One or more employees with special responsibility within research ethics. Their tasks can typically be to:
  - Develop internal procedures that concern research ethics.
  - Stay up to date on research ethics. In particular, to keep in touch with national committees and bodies within research ethics in order to monitor changes in national and international guidelines for research ethics and ensure that the changes are implemented in the organization.
  - Receive reports from whistleblowers about suspicions of deviations from responsible research practice (some organizations will choose to let a manager at a certain level receive these reports).
  - Advise researchers, managers and others on research ethical issues and be resource persons in the internal training in research ethics.
  - Monitor how the organization as a whole functions in terms of research ethics, and report on this to the top management.

  Large organizations typically establish an ethics committee (the name may vary) and have an ethics ombudsman or an ethics officer on their staff. Small organizations make do with one person. These are often experienced researchers with good ethics expertise. Some of them may be working full time with ethics, others part time, depending on the needs.
- An independent, permanent or ad-hoc commission for investigating suspicions of scientific misconduct. The commission's task is usually to make a statement in cases according to instructions from the research organization. The idea behind such a commission is that the most serious violations of responsible research practices should be investigated by independent and neutral experts, so that the suspect as well as any whistleblowers and others involved may receive fair treatment by competent peers. Such commissions are usually composed of persons with the necessary expertise in research, research ethics and law and can consist of internal and/or external members. Some research organizations find it appropriate to join forces with others and establish a joint, permanent commission. In some countries, legislation contains provisions concerning such commissions and the processing of misconduct cases.

Figure 3.1 shows an example of how this can be organized in a somewhat larger research organization:

Comments on the figure: In many research organizations, a leader near the top of the organization is often appointed to be in charge of the investigation in cases of serious violations of research ethics norms (research misconduct). An alternative may be a neutral person outside the chain of command. The nearest leader may be perceived to be too 'close' to the suspect.

In some cases, it may be appropriate or necessary that the independent investigating commission also considers *less serious* violations of research ethics norms. The research organization's internal procedures will indicate when that should

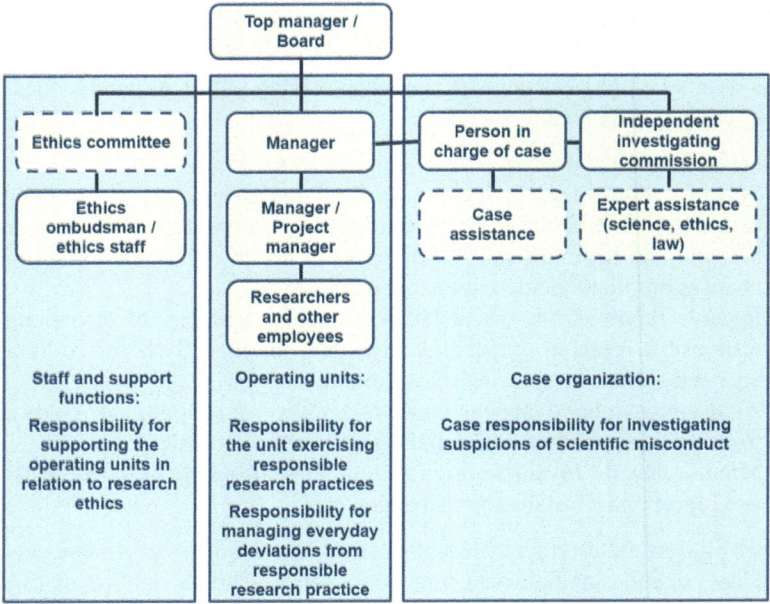

Fig. 3.1 Example of how work with and responsibility for research ethics can be organized in a research organization

happen (this will be further justified in Sect. 23.5.2). However, this possibility is not indicated in the figure above.

References

All European Academies. (2017). *The European code of conduct for research integrity* (revised ed). ALLEA. Retrieved September 21, 2021, from https://allea.org/code-of-conduct/
World Conferences on Research Integrity. (2013). *The Montreal statement on research integrity in cross-boundary research collaborations*. Retrieved September 21, 2021, from https://wcrif.org/guidance/montreal-statement

Chapter 4
The Research Organization's Measures to Ensure Responsible Research Practice

4.1 Ethical Issues Related to Organization and Management

In all organizations, internal administrative, operational and managerial systems will affect how leaders and employees live up to their professional ethical standards. Some systems and schemes promote responsible behaviour, while others can be a breeding ground for irresponsibility. The issues here are diverse, and the connections between cause and effect are often complex – and rarely unambiguous. A comprehensive discussion of this is beyond the scope of this book, but in order to raise awareness of what this is all about, two issues are discussed in the following sections.

4.1.1 The Importance of a High Degree of Professional Morality at the Lowest Level in 'Flat' Organizations

A research organization is a place where tens, hundreds, or thousands of research projects are carried out every year. Most often, each project is carried out by one person or several people in relatively small teams, led by a group leader or project manager. The projects are also generally relatively independent of each other, even when they are connected in programs or are within the same research field. In order for all this to be carried out efficiently in parallel, one must also have support functions to take care of administrative tasks. Some compare the forest of projects to shops in a big department store where the building and its facilities are the common denominator. But that is not exactly the case. Research organizations must have a top management team that is formally responsible for everything the organization does, draws up the main direction of the research activities and takes care of necessary support functions. This requires managerial steering and control. However, the management rarely has dealings with the core activities of a research organization – that is, each research project. The responsibility for and authority over the projects is to a significant degree delegated to the researchers. That it must be so is due to the large number of projects and the fact that planning, implementation and reporting of each project requires an expertise that no manager possesses beyond their own field of specialty. At universities and colleges, the management's distance from the research projects is particularly evident. In these institutions, the academic staff are by tradition and regulations usually given academic freedom that gives them a particularly independent responsibility and great authority. Research organizations are therefore by *nature* what are called 'flat organizations', where the responsibility for the core business (the research projects) is delegated to many at the lowest level (the researchers). Thereby, very few issues related to each individual research project need to go up to or down from upper-level management. *A basic precondition for this form of organization to work, is that all employees have a high degree of professional morality*, so that responsibility and authority that is delegated downwards is managed well and in an ethically sound manner.

In larger research organizations, it is convenient to group the research activities into levels of operational units (often arranged according to research fields), but the division of responsibility between management levels is most often such that in practice one can still talk about a flat organizational model for the research activity. Some research organizations also have other core activities in addition to research (education, for example, in universities and colleges) that may require a different organization and allocation of responsibility.

4.1.2 The Importance of Management and Control Systems that Motivate Researchers

Opinions on how governance and managerial control should be carried out in different organizations, and on which tools to use, are constantly changing. The development is particularly driven by private, owner-controlled companies. The goal is always to find schemes that promote goal orientation, profitability, quality, productivity, efficiency, good conditions relating to health, safety and environment (HSE), etc. New schemes are launched regularly, come into fashion for a while and are implemented by many, only to be replaced by something new. This is reflected in a rich management literature that has long occupied many metres of shelving in major bookstores. Consulting companies, innovative business leaders and 'gurus' often front the development, occasionally based on or subsequently supported by research results. Public organizations tend to follow private companies, as do (public and private) research organizations.

Everyone has something to learn from others, but no arrangement for organization, governance and managerial control is universally applicable to all types of business. The management's overall steering and control of the core activities in a research organization – the forest of research projects led by individual research-ers – must take place in a different way than for the core business of an industrial company, bank or retail chain. Researchers are also generally motivated and demo-tivated in other ways than in the business world. An opportunity to attend a confer-ence; a new and better research instrument; time to complete a book, etc. can motivate more than a salary bonus. Different forms of management by objectives, monitored through a range of performance indicators, which are common in busi-ness, also often have a demotivating effect in research organizations because the indicators (number of publications; number of projects funded in different ways; number of projects on contract for clients, etc.) often do not measure the actual performance – *as researchers see it*. Researchers also have an inherent reluctance to be measured in roughly the same way as a kilo commodity.

When this issue is raised in a book on research ethics, it is because there are examples of management and control systems that indirectly influence employees to prioritize and act in ways that may be ethically problematic. An example, which in recent times has triggered concern among researchers internationally, can illus-trate this:

Example: Traditional management and control systems are based on identifying, setting goals for and measuring mission-critical parameters related to economy, production, etc. The results achieved compared to the targets then form the basis for corporate governance. The research organizations generally end up measuring easily quantified results, such as; number of publications and the like; number of supervised students; number of projects of different categories, etc. These are certainly numbers that are worth monitoring and which in some contexts one can govern by (especially in relation to the research organization's finances), but the problem is that such 'numbers' provide poor descriptions of the real achievements in research. The number of tons of Ariel laundry detergent can be a good measure of production in Proctor & Gamble, because Ariel is Ariel. A common perfor-mance indicator such as the number of scientific publications, on the other hand, gives an uncertain picture of production in a research organization. This is because a publication in a scientific backwater cannot be equated with a publication in a scientific wavefront. And four relatively insignificant publications to which few later refer are not a better but a worse result than an important publication that four hundred later refers to (all may have been published in seemingly equally 'good' journals).

In recent years, many countries and research organizations have used the number of publications as a measure of production, and to a certain extent, researchers and research organizations have been rewarded in relation to how much they publish. This has probably contributed to greater article production in many research groups. Such counting can also be particularly useful in special situations or phases of a research group's development. In well-functioning research environments, such counting often says little. In such environ-ments it is a matter of steering towards the most important research tasks and achieving research results and innovations that matter. The professional ethical concern that many researchers have had about the focus on the number of publications is that it obviously motivates some researchers to choose research tasks and research schemes where the road to publication is particularly short, and to choose ways of publishing that result in as many publications as possible. Some believe that this does not promote academic quality or results of societal importance and that in practice it can easily lead to irresponsible use of society's research resources and unacceptable publishing practices (see Sects. 6.3 and 6.4 and 16.9).

Performance indicators and other forms of statistics are useful and absolutely necessary tools for the management of a research organization. The question, however, is how such data should be used to manage research and motivate researchers and others to perform at their best. The effects may be different than immediately imagined.

The research management's choice of performance indicators will always be significantly influenced by how public authorities rank project applications or evaluate research groups, etc. The number of publications in journals with a high 'Journal Impact Factor' is an example of one such frequently used criterion, until now. However, after a group of researchers in 2012 took the initiative toward the 'San Francisco Declaration on Research Assessment' (Declaration on Research Assessment (DORA), 2012), more emphasis is placed on *the importance and usefulness of research results* than on such simple performance indicators. However, this raises new questions about how one should 'measure' results and performance according to the new criteria in a fair and consistent way in all fields of research, and who is actually qualified to assess research in this manner.

4.2   Measures to Ensure That All Employees Act Morally

As discussed in Sect. 3.3, research organizations have the main responsibility for facilitating that all employees act morally. The tools will be procedures and guidelines for how things must or should be done; resource persons that can guide the individual to act responsibly; leaders who exercise good leadership and have a focus on ethics, and more. Here too, in a book like this, it would be too much to discuss everything, but a small selection of issues and specific measures that are relevant to fostering responsible professional practice in research organizations should be mentioned:

4.2.1   Requirements for Ethical Competence in New Hires and Measures to Maintain Competence

Researchers are by far the most important production factor in a research organization. The hiring of researchers is therefore the managers' most important and difficult task: important because a good choice can be absolutely crucial for the research carried out, while a bad choice can hamper a research area for years, especially at universities and colleges; difficult because it is not easy to find the person who will potentially best fill the position in the short time available to make the choice.

All organizations have employment procedures. At universities and colleges, these are particularly detailed. The main purpose of the procedures is to find the person who will potentially work best in the position. At the same time, the applicants should be treated equally, fairly and objectively. The latter point has led to a significant emphasis on measurable and documented facts about the candidates, such as education, previous jobs, publications and other concrete results, management assignments, positions of trust and other merits that can be read from applications and CVs. It is harder to determine whether the candidates are *skilled* as researchers, leaders, student supervisors, committee members, peers, etc. and what attitudes and personality traits characterize them. This does not appear directly in the CVs, but job interviews, information from references, and various tests can be used to shed light on some of this.

Knowledge of research ethics and proper moral attitudes are elements of competence that applicants must have in order to function well in a research position. Information that can say something about this is rarely found in CVs. When the applications are considered, many employers take it for granted that the candidates have acquired this competence through their education and any previous research jobs. This may be true for most applicants, but experience shows that there is considerable variation in the scope and quality of student courses in research ethics, and the quality of ethics guidance from teachers, managers or colleagues. A fixed item in all job interviews should therefore be to clarify the applicant's ethical competence. The most important thing then is to find out if the applicants base their actions on professional ethics. In practice, it will therefore be necessary both to require a brief description or documentation of all applicants' research ethical training and then, through the job interview, consultations with references and otherwise, assess whether the person has the right knowledge and attitudes in a research ethical context.

A lack of knowledge about research ethics should generally not stop the employment, but the research organization must then ensure that the person in question receives training after employment is confirmed. All research organizations should therefore:

- Have a simple, mandatory programme that provides all new employees, visiting researchers and any students an introduction to the organization's internal procedures and guidelines. The programme may contain a reminder or refresher of relevant national and international guidelines for research ethics.
- Have prepared a full training programme in responsible research practice which is implemented when new employees (researchers and others involved in the research), visiting researchers or others are unable to prove that they are sufficiently qualified in research ethics.
- Have a mentoring programme for new employees where responsible research practice is observed.

While a lack of ethical competence is not automatically disqualifying, clear signs that the applicant has very bad *attitudes to morality* may be grounds for stopping an employment process.

This element in the recruitment process is just as important when a non-researcher, who in one way or another will be involved in the research, is hired. Everyone must be ethically competent. The leaders of a research organization should be particularly well qualified in research ethics because cases involving difficult ethical dilemmas or serious research misconduct always 'flow upwards' in an organization. Leaders come and go, and when new ones are appointed, it is therefore important that the organization has the practice of checking the new leader's research ethical qualifications, in the same way as for new employees. Research ethics should also be a natural element of internal management training.

Age and experience not only grant wisdom, but sometimes also bad habits. Extensive experience is also no guarantee that everything new, will be noticed and taken into account. In addition to ensuring that all new employees and leaders are suitably qualified in research ethics for the position, one must therefore also ensure that competence is maintained. The organization should have measures to ensure this happen. Some will immediately think of courses and other concrete measures. However, experienced researchers are rarely motivated to attend refresher courses in research ethics. In practice, therefore, it should perhaps be thought more in the direction of establishing a work culture where the experienced as well as the inexperienced help each other to discover and correct even small deviations from responsible conduct, without making a big issue of it. Attention to the small deviations will be a constant reminder of ethical issues, create positive attitudes and reduce the possibility of developing bad habits. The responsibility for creating a working environment where this happens lies primarily with the managers at the lowest level of the organization – they are closest to it. Another opportunity to correct employees who appear to be developing in the wrong direction or who do not take into account developments in research ethics is annual employee conversations between employees and their nearest managers.

4.2.2   Quality Assurance and Control Systems that Include Research Ethical Issues

One of the most important arguments for investing in research is that the knowledge it produces is reliable and of high quality. Systems and working methods that ensure high quality are therefore a central part of what one might call 'the scientific method', but are not unique to research. Many critical work tasks in the business and public sector are, for example, subject to equally good or better quality assurance and control than is usual in research.

In everyday life, *quality* is a concept used in various ways and in many contexts. One way is to look at quality as a term for compliance with expectations and requirements of the work or the results. By *high quality*, one will then mean the *right quality*, that is, a high degree of compliance, while poor quality means that the work or the results have critical shortcomings in relation to expectations and requirements. This way of defining quality is used in the most common international standard for quality management systems, the ISO-9000 family of standards (International

Organization for Standardization (ISO), 2015). On this basis, a 'high quality' research work will presuppose both:

- High scientific quality: The research topic is relevant and important (in a scientific or societal context); the results are accurate and clearly described; the premises for the results are carefully explained; the possible sources of uncertainty and error are described and taken into account; the conclusions are made on a factual basis, etc.
- High quality in other contexts: The financial support contract and cooperation agreements are respected; the research project follows the time and cost plans that have been set; the work is carried out in accordance with relevant guidelines for research ethics, etc.

With such a definition of quality, professional ethics becomes a natural element in the concept of quality: Everyone expects or demands that researchers follow relevant research ethics norms, and poorly justified deviations from them are considered poor quality, as with poor scientific quality. The quality system in a research organization must therefore contain elements that ensure everyone is working in accordance with responsible research practices.

Quality assurance and control can be exercised in many ways, and different organizations choose different solutions. In research organizations, however, there are two instruments for quality control that are used by everyone:

One is the tradition that it is first and foremost the researchers themselves who must control their own work. Checking one's own results, and checking again, is deeply rooted in the work tradition of researchers. Researchers take pride in being accurate, and know that the loss of reputation resulting from being caught being sloppy and making mistakes one should have seen, is great. However, checking one's own work has its own weaknesses. It is, for example, easy to become blind to professional mistakes and shortcomings in one's own work. Others can often more easily see this. Quality control based on researchers checking their own work is also unreliable when it comes to ensuring that the work follows good research ethical practice, since many violations of ethical norms are result from ignorance or misconceptions, and some researchers deliberately cheat. A credible quality system in a research organization must therefore contain more instruments than this.

The second instrument for quality control is the tradition for self-monitoring to be supplemented by an *external peer review carried out by a publisher or a conference organizer in connection with the publication of the research work*. It is obvious that such peer reviews contribute to quality, but they also have a number of weaknesses that are discussed in more detail in Sect. 19.2. It is also a problem that this quality control varies considerably in scope and thoroughness. The situation has become particularly obvious after the emergence of thousands of new online journals operating according to a different business model than the traditional, subscription-based journals (more on this in Sect. 16.9). In addition, many research results are published in a different way than in peer-reviewed journals and conferences, and for various reasons some results can not be published at all (examples are discussed in Sects. 9.4.4 and 18.3). This is not least true in the institute sector and industrial R&D units, where research reports targeting clients and users can account for the bulk of the production.

Publication means that other researchers can gain insight into the results and can compare them with their own work and that of others. This helps to detect errors and wrongdoings, and a number of cases of research misconduct have subsequently been uncovered this way. However, this cannot be called quality assurance, but rather a delayed and indirect quality control that sometimes takes place long after the error or misconduct has been made, but which can limit further damage caused by the wrongdoing.

Because external peer review under the auspices of publishers and others only covers part of the production in a research organization (this varies greatly from organization to organization) and due to uncertainty about how reliable some external peer reviews are, research organizations should also have *internal systems* for quality assurance and control of research projects, which then come in addition to the two traditional forms of quality control described above. The justification for such a system is further strengthened by the fact that violations of research ethics norms are not uncommon (see Chap. 21).

For research institutes with a large portfolio of projects on commission for external clients, such systems are often a prerequisite for acquiring projects, with the application of the systems having to be documented in each project. The challenge with such systems, however, is not putting them in place but in making them work in practice; see later. Such systems also do not give any guarantee that absolutely everything wrong is detected, because the scope and thoroughness of any quality control must be adapted to the organization's resources and operating conditions.

Each research organization may, with some exceptions, decide for itself how such an internal quality system should be.[1] However, there are international standards for quality management (International Organization for Standardization, 2015) that can be used as a framework for quality assurance and control of all kinds of activities in all types of organizations, which many research organizations therefore use. A quality management system can be established at several levels in an organization. Most have one that embraces the entire organization, but it is also possible to have specific quality management systems for underlying units, laboratories, teams, individual projects, special activity areas, etc. In addition, all organizations have systems for 'internal control', i.e. systematic measures to ensure and verify that the organization operates according to set rules, policies and procedures in specific areas such as accounting and auditing, compliance with laws and regulations, etc. The implementation of certain internal control systems is statutory. Internal control systems can to some extent be seen as specific elements in the organization's overall quality system.

The overall quality system in an organization is often developed by the administrative staff and top-level management. It often contains layer upon layer of procedures for everything that has to be performed in a systematic way in the organization.

---

[1] In some areas, national legislation may have provisions that organizations must have a system for internal quality assurance or internal control of special tasks or obligations related to research.

The majority of it is administrative procedures, quite removed from the researchers' world of immersion in documents, experiments, observations and theories. Researchers see these quality systems as useful and necessary but somewhat bureaucratic regulations. Much is perceived as the systematization of the obvious, and general popularizations, such as 'our ten commandments for quality', hardly create long-term motivation. In some organizations, there is also far more emphasis on getting the system in place and making it known than on the follow-up of how it works in practice. An internal procedure requiring all employees to follow relevant guidelines for research ethics, and a course in research ethics for new employees, is a necessary though insufficient measure to ensure responsible conduct of research. Guidelines are most often something researchers read once and then occasionally look up. A course is also a one-time event. There is soon a loss of focus on the content of both measures in the researchers' everyday lives.

An internal system for quality assurance and control of *research works* will usually include two measures: A well-qualified researcher in the organization (or one brought in from outside) *checks or verifies* the work, and a responsible manager *approves* it. Their identity is made known. There must also be a description indicating the scope and procedure for the verification and approval work. The approval is first and foremost a confirmation that the organization believes that the project obligations have been fulfilled, that the work is in accordance with responsible research practice, and that it is quality assured. Most often, verification and approval are carried out when the work has been completed, but before it is reported or published.

If necessary, an internal quality system can be set up so that the internal peers can go deeper into the research work than the publishers' peer reviewers usually have the opportunity to do (more about peer review in Sect. 19.2). Researchers who review the work of other researchers in their own organization cannot be said to be neutral. But at the same time, they also have a personal incentive to be critical and accurate – letting work with poor quality pass weakens the reputation of their own organization. That benefits no one. The prerequisite is that those who are there to verify and approve the research work of others have *time for the task and the right knowledge and attitude for the work*:

- Verification and approval requires that internal peers and leaders work conscientiously with the task and take the time required, based on the prerequisites of the assignment. However, lack of time in the final phase of a research project often causes the researcher to push for a quick verification and approval ('the article/report needs to be sent tomorrow'). Inadequate funding and lack of personnel with available time may also jeopardize the system.
- If the quality system is also to ensure that guidelines for research ethics are followed, the person reviewing the work must be qualified for it. Several cases of scientific misconduct indicate that even senior researchers and managers may have insufficient ethical qualifications or weak ethical judgement.
- Getting work reviewed by colleagues should be seen as a great help toward ensuring high quality. The transparency inherent in this is both one of the most important principles in science and a sign that there is nothing to hide. On the

other hand, researchers usually want to show that they have 'full confidence' in each other. Quality control of the work of a colleague can easily be seen as the opposite of this, because all control is also based in the possibility that something may be wrong. Many may also be afraid of getting into difficult situations. For example, it is easy to imagine that a researcher who, during the review of a colleague's work, discovers examples of biases that may be related to the colleague's political views or personal interests, may end up in an unpleasant discussion when this is addressed. Younger researchers may also be reluctant to quality control an older colleague for fear of unpleasant consequences if they find anything questionable. The working environment in the research group is decisive for how this is experienced.

• Although everyone knows that mistakes, undesirable practices and worse things happen from time to time, very few think that it can also occur in one's own workplace. The desire to show confidence in colleagues may also lead some researchers to try to avoid having to control colleagues' work. This is somewhat inconsistent. As an external peer reviewer, most researchers have few concerns about controlling the work of other researchers and providing critical commentary. However, the fact that this often happens anonymously may come into play.

Management's focus on the work of the internal peers who are engaged in internal quality assurance and control is therefore highly important.

Modern research very often takes place in teams of two or more working together on a project or working on different but closely related projects. A group of researchers is likely to discover errors and irresponsibility more easily than one researcher alone, and the likelihood that several researchers simultaneously behave ethically irresponsibly is less than the likelihood that one researcher does so alone. A quality system in a research organization may take advantage of this by focusing on *the individual research team* and providing procedures and guidelines for how the team should work together openly and take joint responsibility for the quality and ethical issues of their own work. By giving the research team a central place in the internal quality assurance system, it is implicit that in environments where the researchers have a habit of isolating themselves or working alone, something special should be done to establish local research communities. A team-focused quality system could place more emphasis on attitudes than on procedures. It should focus on creating teams where it feels natural to have insight into each other's work, where it is easy to address difficult issues and where everyone understands the importance of both helping and controlling each other. The team leaders (group leaders, project managers, student supervisors and others) have the lead role in this. Their leadership skills and personality are essential for this type of quality management system to work. The main task and responsibility of higher-level leaders will be to monitor how the team leaders function and contribute with supportive measures to help them perform well.

A team-focused quality management system can be realized in many ways. The operating conditions for the organization, the research profile, the composition of the staff, etc. will determine which solution should be chosen. These are topics that go far beyond the scope of this book.

4.2.3   *Working Environments that Promote Responsible Research Practice*

The individual researchers' personality and character traits form the basis for how they relate to ethical norms. For many, unethical behaviour is unthinkable, while some are more easily tempted to act unethically. However, the work environment and circumstances around it also play a significant role in how people act. A 'bad' working environment or 'unfortunate' circumstances can become a catalyst for acting irresponsibly. Even a researcher who in principle cannot imagine doing something wrong may then be duped or tempted into doing so. There can be many reasons for this:

- One obvious reason why someone breaks with responsible research practice is that they somehow benefit from it, and that the benefits are perceived as far greater than the likelihood or consequences of being discovered. Researchers become tempted or feel pressured to cheat in order to make the research results publishable; to get more, or more sensational, publications; to improve an application for research funding; to gain an academic position or a prestigious honour, etc. Strong individual competition between colleagues can promote this. The degree of intent can vary greatly.
- Pressure from managers or clients can also trigger wrongdoing. One example can be a project manager who forces an employee to keep quiet about scientific misconduct by another researcher in the team. Another example can be a client who threatens a researcher not to report results that are negative for the client.
- Even reasonable demands and expectations from colleagues, management, clients and others can be stressful in certain circumstances, especially if the researcher does not feel capable or able to meet the expectations. Instead of addressing the issue, the researcher feels pressured to meet the requirements and expectations by cheating.
- In research environments where some researchers are not so careful about following good practice, honest researchers can also become 'infected' and start acting improperly. When boundaries are pushed and broken by others in the group, it is easier to start interpreting the research ethics norms liberally and believe that a little cheating 'is not so dangerous'. This is rooted in people's inherent urge not to stand out from the crowd. When others benefit from doing wrong, it becomes especially difficult to resist the temptation to do the same. It is also easy to imagine that in working environments where researchers do not mind exaggerating a little in project applications, playing around somewhat with project funds, 'decorating' research results, etc., the path to more serious breaches with good research practice may be a short one.
- In research environments where ethics are not part of everyday life, there is a danger that research ethics guidelines gradually become forgotten or regarded as less important. Everyday training in distinguishing right from wrong and exercising sound ethical discretion will also be diminished. Deviations from respon-

sible research practice may then be committed as a result of negligence or ignorance. Inadequate or absent guidance from professors, mentors and leaders is another aspect of this.

- When investigating cases in which researchers raise suspicions that colleagues have violated research ethics norms, it is not uncommon to find an underlying conflict between the whistleblower and the suspect. It may, for example, be colleagues, partners, student/supervisor, researcher/client, etc. who initially worked well together until something happened which the parties or one of the parties were unable to find a solution to. Step by step, the conflict grew, the solution-oriented dialogue between the parties stalled and was replaced by barrages of accusations from both sides. Researchers who despair of not being heard can resort to self-willed actions – for instance, publishing joint works only in one's own name, not referring to or acknowledging other parties contributions, going on alone with project concepts originally shared with others, etc. The devastating moment in these cases often comes if one of the parties raises one of the sharpest weapons a researcher has: accusing others of research misconduct (often in the guise of being a whistleblower). Experience from the investigations that follow in these cases often shows that the accusations are exaggerated and that the person who made the accusations has also acted irresponsibly. In several cases it has also been found that the management has in one way or another acted incorrectly or failed. Conflicts in the workplace are therefore a potential source of breaks from responsible research practice – a warning sign that the organization's management must respond to as early as possible.
- Violations of research ethics norms can probably also be triggered by laziness or poor work motivation. Rather than completing a time-consuming interview survey, the researcher fabricates some interviews. Instead of staying in the laboratory throughout the evening to complete a long series of measurements, the researcher draws in some extra measuring points that look plausible – and then goes home for dinner.

The review above is to a large extent based on Anderson et al. (2013), Drivdal et al. (2019) and Gopalakrishna et al. (2021).

Guidelines for research ethics and good initial training in research ethics will probably only have a limited impact on whether researchers act irresponsibly for the type of reasons discussed in the examples above. The problems here lie in the organizational culture and the working environment. That makes it all difficult. Establishing procedures, guidelines and ethics courses is relatively easy. Developing operating conditions and creating a working environment that reduces the need, desire and opportunities for sloppiness and cheating is far more demanding. There is no simple and universal recipe for how a well-functioning research environment should be realized in this context. However, it is reasonable to assume that immorality will have a bad breeding ground in research environments characterized by openness and community, where it is easy to talk about things, where one daily feels colleagues' zero tolerance for breaches of responsible research practice and where everyone takes pride in acting ethically and legally. Senior managers and leaders who are good role models and appear neutral (who do not favour or exhibit bias) and react quickly when situations that can lead to conflicts arise will also be important prerequisites. A team-focused quality assurance system, as outlined above, can be one element in facilitating such a work environment.

4.2.4   Goals and Strategies that Researchers Support and that Inspire

From time to time, differing opinions about the prioritization of research areas and research projects can lead to internal conflicts in a research organization, often between researchers and leaders. As discussed above, such conflicts can in turn trigger ethically irresponsible behaviour. One measure that can prevent and limit such events is to develop goals and strategies at different levels in the organization that employees support and find inspiring.

The strategies at the top level and large subunits are often developed in the same way as in the business sector, perhaps also facilitated by external consultants. They often result in strategy documents with big visions and goals. These processes are especially useful for the organization's top management in that everyone's eyes are aimed at the future. Senior managers, who may not have as much cooperation on a daily basis, thereby experience something that unites them, something to reach for together. Researchers often feel alienated in such processes and are only moderately interested in overarching visions and goals. They are concerned with strategies related to finding molecules with unexpected properties or documents that provide a new understanding of important historical events.

The strategy work in a research organization should be adapted to the organization's uniqueness and operating conditions. The strategy *processes* then often appear to be more important than the strategy documents. In a research organization, a good strategy process must first and foremost include the issues that are most relevant to the *researchers*. It must be open and involve and engage all. The process that shall lead to the organization's overall strategy should then focus on creating an understanding of the organization's operating conditions and societal responsibilities. The processes at lower levels should focus on selecting directions in research that excite and are perceived as important and useful, scientifically and societally, and which in addition contribute to the realization of the overall strategy. For the individual researchers the process should also help them to open their eyes to the importance of others' research and to see the strength in collaborating with others. For the leaders the process should contribute to a wider understanding of the opportunities in each field of research and the potential the organization has in developing both the breadth and the depth of the research. At all levels, the strategy processes should contribute to creating the greatest possible support for the basis of priorities and management at the various levels.

4.2.5   *Management Focus on* Implementing *Responsible Research Practices*

Investigations into serious misconduct cases often show that the research organization where the wrongdoing has occurred can also be criticized. An example of this which provides a good illustration of parts of the research organizations' responsibilities is given in the box below.

An Example of the Expectations Towards a Research Organization When It Comes to Facilitating and Ensuring that Employees Act Morally
In Norway in 2006, an independent commission was appointed to investigate a case of suspected research misconduct. The case concerned a university hospital researcher who had fabricated data to a large extent and over a long period. In addition to 'convicting' the researcher, the commission also directed strong criticism at the research institution (Ekbom et al., 2006, translated from Norwegian by the author):

> When a research institution, as the hospital is, facilitates research at its institution, it must be prepared to bear full responsibility for the individual researcher and the relevant research project, regardless of whether others also have an independent responsibility. Patients and others who relate to the hospital, including cooperating institutions, must be able to assume that researchers at the hospital operate on behalf of the hospital in question, and that the hospital has the overall responsibility.
>
> The health institution must thus take criticism for what appears to be inadequate training, management and control of [named researcher] and other employees' research activities at the institution. This has probably been a contributing reason why the fraudulent research was able to take place and continue over such a long time.

The criticizable conditions can be summed up as follows:

- Lack of initial control and organization of the doctoral project of [named researcher], including description of responsibilities.
- Lack of training and failure to make [named researcher] and other employees aware of rules for handling patient material, pre-assessment of research projects and authorship.
- Inadequate management and procedures for discovering and handling deviations from internal instructions etc. (p. 114)

The criticism was directed at the institution's top management.

Despite the fact that many investigations into individual researchers also end up with criticism of the research organizations' management, as in the example above, it rarely has consequences for anyone. When the research organization's management and board do not adequately ensure that the research is carried out according to recognized research ethics norms, they violate both national and international guidelines for research ethics and in some cases perhaps even national laws and regulations. The research organizations themselves, and perhaps others, should therefore react and sanction this to a greater extent than they do today.

In recent times, 'ethics' has become an area both boards and top managements have begun to grasp. Most organizations therefore have procedures and overall codes of conduct in place and do much to raise awareness of the importance of ethical accountability. Many organizations also put much effort into presenting this to employees, public authorities, clients, partners, etc. Lectures on ethics are held at employee gatherings and statements on ethics are posted on the organization's website. However, most people are more moral in thought than in deed, and in many organizations there can easily grow a gap between the work on ethics at the top level in the organization and the daily, ethical management and quality assurance work at the bottom of the organization aimed at ensuring that responsibility and good morality characterize every single research project. The research organization's responsibilities include both systems and practices, but it is the ability to *implement* the systems in a natural way in the daily research work which in an ethical context probably separates good research organizations from bad.

References

Anderson, M. S., Shaw, M. A., Steneck, N. H., Konkle, E., & Kamata, T. (2013). Research integrity and misconduct in the academic profession. In M. B. Paulsen (Ed.), *Higher education: Handbook of theory and research* (Vol. 28, pp. 229–235, 235–241). Springer.

Declaration on Research Assessment. (2012). *The San Francisco declaration on research assessment.* Retrieved September 21, 2021, from https://sfdora.org/read/

Drivdal, L., Kaiser, M., Hjellbrekke, J., Rekdal, O. B., Skramstad, H., Torp, I. S. & Ingierd, H. (2019). *Forskningsintegritet i kontekst: Resultater fra en kvalitativ studie.* [Research integrity in context: Results from a qualitative study. 3rd sub-report from the working group in the research project RINO (Research integrity in Norway). In Norwegian]. Norway. University of Bergen, The Norwegian National Research Ethics Committees and Western Norway University of Applied Sciences. Retrieved September 21, 2021, from https://www.uib.no/sites/w3.uib.no/files/attachments/rino_3_-_forskningsintegritet_i_kontekst_-_resultater_fra_en_kvalitativ_studie.pdf

Ekbom, A., Helgesen, G. E. M., Lunde, T., Tvedal, Aa. & Vollset, S. E. (2006). *Rapport fra granskningskommisjon oppnevnt av Rikshospitalet – Radiumhospitalet HF og Universitetet i Oslo 18. januar 2006.* [Report from the Commission of investigation appointed by Rikshospitalet-Radiumhospitalet HF and the University of Oslo on January 18, 2006. Issued Juni 30, 2006. In Norwegian]. Retrieved September 21, 2021, from https://jusboka.no/wp-content/uploads/2016/07/Rapport-fra-granskningskommisjon-oppnevnt-av-Rikshospitalet-Radiumhospitalet-HF-og-Universitetet-i-Oslo-18.-januar-2006.pdf?x22677

Gopalakrishna, G., ter Riet, G., Cruyff, M., Vink, G., Stoop, I., Wicherts, J. M. & Bouter, L. M. (2021). *Prevalence of questionable research practices, research misconduct and their potential explanatory factors: a survey among academic researchers in The Netherlands.* MetaArXiv Preprints. Retrieved September 21, 2021, from https://osf.io/preprints/metaarxiv/vk9yt/

International Organization for Standardization. (2015): *ISO 9000 Family – Quality management.* Retrieved September 21, 2021, from https://www.iso.org/iso-9001-quality-management.html

Part II
Morality and Responsible Practice in Daily Research Work

Chapter 5
The Researcher in the Workplace: Employee Morale

5.1 Employee Morale

At work, researchers must first of all live up to society's general perceptions of good and bad behaviour in the workplace – i.e. the common norms of employee morality.[1] Most will then associate *good employee morale* with employees who show positive attitudes towards work, the workplace and their colleagues. Who find joy in doing what they are supposed to and often go above and beyond their formal duties. Who deliver high-quality work, actively participate in creating a positive work environment, take collective responsibility and show a healthy loyalty to the workplace. *Poor employee morale* manifests itself in low work effort, cheating, sloppiness, disloyalty, bad behaviour towards colleagues and managers, etc. Good and bad employee morale can appear in slightly different ways in different professions.

Some of the general principles for how employees should behave at a workplace (their rights and responsibilities) are established in law. The reader should make themself familiar with this in relevant national laws. All organizations also have internal policies, procedures and codes of conduct that are relevant to many ethically related issues in a workplace. The working contract and any agreements with employee unions may also have relevant provisions here. Employees therefore have a good deal to relate to in order to function well and ethically responsible in a workplace.

---

[1] The opinions on right and wrong behaviour in a workplace have changed in time and may be different from culture to culture, depending on society's view of 'work'. In organizations and enterprises with a modern view of work, the interests, rights and responsibilities of employer and employees are well balanced. This is mirrored in the employee morality in research organizations today. Matters related to the workplace, employers and employees are also established in law in many countries, and in many contexts legislation and ethics will then go hand in hand. The description in this book is to some extent based on conditions in Europe, but are probably just as relevant in many other countries as well.

© The Author(s), under exclusive license to Springer Nature Switzerland AG 2023
D. Slotfeldt-Ellingsen, *Professional Ethics for Research and Development Activities*, https://doi.org/10.1007/978-3-031-25484-0_5

Rather than addressing the full range of issues related to ethically responsible behavior in a workplace, the discussion below is limited to issues that are particularly relevant in a research organization. Most of what is said can be perceived as obvious and banal, but has to do with situations that everyone in a research organization may have problems with from time to time.

5.2   Common Decency in the Workplace: The Relationship with Colleagues, Managers and Others

The workplace, as a social and professional meeting place, is a main arena where the individual researcher's morality and ethical assessments are tested daily. This concerns ordinary things such as:

- What thoughts we have about others and ourselves: Are we tolerant or prejudiced toward others? Do we have a nuanced view of both ourselves and others or do we only see our own greatness and the mediocrity of others?
- How we speak or write to/about other people: Are we reasonable or unreasonable, respectful or condescending, attentive or self-absorbed, open or intriguing, considerate or rude, solution-oriented or combative, fair or slanderous and gossipy, etc.? When considering whether a statement about another person is respectful, one must first ask whether the facts are correct, whether the allegations or assumptions made can be substantiated, and whether the words and arguments are justified (for example, free of personal attacks, affrontive personal descriptions, or speculation about the person's motives). When speaking to and about others, both the words and the context are of importance (said with calmness or aggressiveness, said once or repeated constantly, in confidence or in public, etc.). One must also consider whether the chosen form of expression is justified and necessary as the case stands. A normal approach will, of course, be to speak directly to those affected, or at least internally within the organization. To spread opinions about others on social media or in emails to many is rarely necessary.
- How we collaborate with others: Are we inclusive or discriminatory? Do we argue for what we want to achieve, or do we use force to make it happen? Is it important to lead and control? Are we creating alliances *for* something or *against* something?
- How we compete or disagree with other people: Do we fight openly or do we create intrigue? Are we attacking the opponents' arguments, or them as persons? Do we focus on what we agree on or what we disagree on?

Most people develop a good understanding of right and wrong in questions like this during their upbringing and basic education – common good behaviour, essentially. Employees in a research organization are usually positive toward their workplace and experience good relations with their leaders and co-workers. However, leaders and employees exhibiting unacceptable behaviour towards others also make appearances in research organizations, and from time to time and in the worst cases can destroy a good work environment. One person who behaves badly towards others can be enough for that to happen.

In research ethics, good behaviour is, among other things, rooted in norms of respect for others. Having respect for other people does not mean agreeing with them on everything, or being supportive and tolerant of their opinions and actions. Respect is treating others in an objective, open and considerate way. Everyone can, in a moment of carelessness, in a stressful situation, etc. come to act disrespectfully towards others. It should be easy to make amends, if necessary. Occasionally, however, one also meets people whose lack of respect for others seems to have become a part of their character.

Disrespect towards others in the workplace can take many forms: Disparagement of other people's work effort or qualifications; offending remarks; casting suspicion on other people's motives; hurtful joking, teasing and scolding; malicious inducement of stress; threats; inappropriate sexual attention, etc. To ignore, isolate, freeze out, fail to inform, sabotage, slander and spread rumours about others is also disrespectful behaviour. Disrespectful actions can be a violation of several research ethics norms, such as truthfulness, objectivity, openness, open-mindedness, obedience of the law, and more. Everyone in the workplace can, through their own actions and by reacting to others, contribute to disrespect not being tolerated in the work environment. Otherwise, of course, this is also among the leaders' foremost concerns.

The severity of disrespectful actions towards others increases significantly if they are repeated towards the same person or persons several times over a longer period of time, and especially if the victims are weaker than those who attack, or have fewer opportunities or abilities to defend themselves. The latter also applies if essentially resourceful individuals become so tormented that their ability to defend themselves is impaired. All people, including leaders and other people in positions of power, can be hurt so strongly that it affects their health and welfare. When disrespect goes so far, it is a matter of harassment or bullying, and one moves seamlessly from ethical assessments of what is good and bad behavior in the workplace to legal assessments of what is a breach of the law.

5.3 Bullying and Harassment

If a person is subjected to serious and unwanted abusive, intimidating, hostile, degrading or humiliating treatment from one or more other persons, it is a case of harassment or bullying. Many people are bullied by more than one person at a time, and sometimes the bullying is directed at more than one person. The state of understanding on bullying and harassment in the workplace and a comprehensive insight into the international research in this field is given in Einarsen et al. (2020). Harassment and bullying in the workplace are offences in many countries, and more countries will eventually establish legislation in this field. The ILO Convention No. 190 on the Elimination of Violence and Harassment in the Workplace (International Labour Organization (ILO, an UN-agency), 2019) will contribute toward this.

The threshold for what is termed harassment (both in an ethical and legal sense) changes over time. Various forms of sexual attention that society silently accepted a few years ago are now prosecuted as sexual harassment. Defamation that could have been prosecuted a few years ago can today be posted on social media or elsewhere on the internet with virtually no consequences, and mobile phone bullying is a whole new problem ('cyberbullying'; see for instance Vranjes et al. (2020)).

When assessing the severity of disrespectful behaviour, one must not only consider the action or expression itself, but also take into account how problematically the injured party views the behaviour and how it affects the person's work situation, health, welfare, functioning, safety, etc. The line drawn by management between what is permissible and impermissible is largely obscured. However, a dismissal case at the University of Oslo in 2009 which ended up in the court system (harassment is illegal in Norway) can be taken as an example of the thinking around this. The arguments here should be relevant in other countries and subject to other legislation as well. The case is interesting also because the researcher's freedom of expression was weighed against the consequences of harassment for the working environment at the institute. The case, which is discussed in the box below, sparked much debate.

> **An Example of *Freedom of Expression* in Academic Institutions Being Strongly Protected, but that the *Form of Expression* Must Not Conflict with Commonly Accepted Norms**
>
> In 2009, a professor at the University of Oslo was dismissed for gross violations of his duties through improper utterances about others as well as insubordination. The validity of the dismissal was tested before the District Court and the Court of Appeal, which ruled in favour of the university. The Supreme Court's appeals committee declined to consider the case in the Supreme Court as it found no prospect that the ruling would be changed. The case was unique in the university's 200-year history.
>
> A starting point for the case was that the professor believed that his field was downgraded by the institute. His main concern was to speak up and fight against it. The problem was the way he argued for his case. Instead of sticking to facts and academic reasoning, he attacked the department head as a person.
>
> He wrote, for example (judgment in Oslo tingrett [Oslo City Court], 29.1.2010, case 09-043651TVI-OTIR/05, translated from Norwegian by the auther): '… [She] definitely has a problem with academic weight, and her professional field of interest is very narrow. On the other hand, she has a strongly developed sense of solidarity to the rest of the West End clique she is part of. Those who pushed for her election were precisely her West End

(continued)

friends who thought in this way they could secure their interests in the allocation of resources and power at the institute.'

Statements of this kind, where one speculates about other people's motives
and 'plays the man and not the ball', are not uncommon in everyday discussions. However, it is worse and more unusual when it happens in writing. As
a single statement, perhaps made in anger or despair, this may perhaps be
something one has to endure in a workplace. On the other hand, such utterances violate several research ethical principles such as fairness, that the criticism should be constructive, and that one should have respect for other
scholars. In this case, it also turned out that some of the statements were
untrue or inaccurate. Despite this, an apology for the wording could possibly
have ended the case. The problem, however, was that the professor continued
to make arguments of the same nature even after leaders and colleagues had
reacted to them, and he distributed his statements in emails to many others
internally and externally. He had also made unsubstantiated and offensive
statements against others and in other contexts as well. The court therefore
had no difficulty in describing the professor's statements as 'grossly offensive, as the form of expression is neither factual nor necessary to convey the
message.'

In its argument the university emphasized that the problem was not that the
professor had stood up for his case, but his course of action and the consequences it had for the working environment at the department. Many, however, supported the professor. In part, they downplayed the emphasis on his
choice of words and his insubordination. They saw the case as a conflict
between the leadership of the university and a professor who used his freedom
of speech to fight for better conditions for his academic field. In part, they
disagreed that dismissal was an appropriate sanction in this matter. This was
therefore a case that also dealt with the boundaries of the research organization's rights to govern and apply forms of sanction (see Sects. 23.1 and 24.1).

The conclusion to this case was not drawn by a unified research community, but by the judiciary. It is thus also an example of how common societal
values and norms (on which legislation is based) apply in the research
community.

The extent of perceived harassment and bullying in the workplace will vary from
organization to organization. National rights for workers, working traditions and
legislation will affect this. The extent of harassment and bullying is regularly charted
in many countries, and many organizations have routines for regular internal work
environment surveys. Two examples from Norway, a fairly transparant and civilized
country, can illustrate what one might find: In a survey (Statistisk sentralbyrå,
2017), 8% of the employed in Norway stated that they had been involved in uncom-
fortable conflicts with superiors at the workplace, often or sometimes; 8% ditto with

colleagues; 2% had been exposed to harrassment or teasing by colleagues, a few times a month or more; 2% ditto by superiors; 4% (men 2%, women 7%) had been exposed to unwanted sexual attention, remarks etc., a few times a month or more. In a survey on bullying and harassment among employees in the university and college sector in Norway (Ipsos, 2019), 13% of employees responded that they had been bullied or harassed at their current workplace in the last 12 months; 1.6% of women and 0.7% of men had experienced sexual harassment (one or more times) during the same period. However, readers should inform themselves about the situation in their own organization.

Bullying can ruin a working environment and cause victims to suffer and become sick. Bullying has therefore gradually been seen as a societal problem that we should get rid of. The #MeToo movement, which started in the US film industry in 2017 and spread around the world, focusing on sexual harassment, has given new pace to dealing with the problem, also in the research community.

5.4   The Danger of Developing Prejudice and Fixed Ideas

Most people adopt certain preconceived and unreasonable opinions about this and that. This also manifests itself from time to time in researchers, who should know better – prejudice is not compatible with science. Prejudices often emerge in challenging situations, for example in public debates, where many may have witnessed researchers and other professionals expressing negative and biased opinions about the relevance and quality of their opponent's arguments, or their professional competence and motives. Prejudice makes it difficult to have a constructive scientific discourse and to agree on facts, priorities and uncertainties. This weakens people's trust in researchers and reduces the credibility of research-based knowledge as a basis for decision making in society. It is also not uncommon for employees to believe that managers, clients and other stakeholders have opinions and motives that they often do not hold at all. Many research organizations have experienced how this can lead to conflicts and personnel problems.

Holding prejudices is probably part of human nature. However, one of the most important elements of the formation process needed to become a good researcher is to train the ability to think factually, critically and analytically. The ability to rid oneself of prejudice in a professional context is central to this.

Holding fixed ideas in relation to scientific questions has something in common with prejudice. Researchers can become so fixated on their own subjects, methods, theories, research results, etc. that they are blind to or disinterested in alternative views and results. The similarity with prejudice lies in the fact that whoever holds fixed ideas often rejects or has an impaired ability to conduct constructive discussions with other professionals.

Prejudice and fixed ideas are examples of attitudes and habits that establish themselves gradually, most often without the persons themselves realizing that it is happening. Others, colleagues and leaders, see this more easily and therefore, in clear terms, can make a constructive contribution towards correction. The earlier one is corrected by colleagues and others, the easier it is to change behaviour. Here lies the great dilemma in all workplaces, to balance between having tolerance for the peculiarities of others while reacting if they behave inappropriately.

5.5   Working Hours and Presence in the Workplace

Researchers mostly work very much. This is partly due to the fact that the concentration, dedication and excitement of doing research easily leads to different degrees of obsession – one has difficulty letting go of the tasks. This is also because the subconscious seems to play an important role in the thought process one has to go through in order to develop new knowledge. Facts and theories must lie there and ferment, all the time and over long periods, before ideas and understanding form in the consciousness. Therefore, one often has to live with the research in order for it to produce results. Researchers' inherently positive attitude to their projects and research fields and the great work effort most researchers make are hallmarks of a good work ethic.

But even if researchers usually work a lot, excessive *absence from the workplace* (by working at home for long periods, constantly attending conferences, etc.) can from time to time become a problem. Although modern technology makes it possible to perform a number of research tasks without being in the workplace, the main rule is usually that employees should be present in the workplace during working hours, according to specified organizational rules. Absence should generally be well justified. However, many research organizations do what they can to allow researchers the freedom to adapt their workday to suit their needs and wishes, especially at universities where the rules are also often practised liberally. However, in order to demonstrate good employee morale in a research workplace, one should use this freedom responsibly and considerately, i.e. one should not only do what is best for oneself but *also take into account what is best for others*:

- Sometimes participants in a research project work far apart, other times side by side in the same workplace. In the latter case, work should be organized so that the efficiency and creativity of the team is optimized. That may require a presence in the workplace at times other than what is best for each team member individually.
- Contributing to the development of the research environment in the workplace can be almost as important as the research itself. It requires presence.
- One of the central elements in the Humboldtian university model, which has influenced universities worldwide, is that there should be close contact between professor and student. Students shall not only learn, but also understand and develop as human beings. Presence at the university, along with the students, is a prerequisite for that. Modern technology – email, video calls, webinars, social

media, etc. – now offer universities new opportunities to enhance the contact between professor and student even when they are away from each other.

- The feeling that 'everyone is equal' has great value in modern democracies. If researchers gain or give themselves 'privileges' in the form of strongly deviant work arrangements, excessive absence through travel, etc., other employees who have to follow standard working hours can easily become both demotivated and provoked. As an element of exercising good employee morale, researchers should take into account such psychological effects when taking advantage of their special freedoms.

The freedom to control one's own time is greatest at universities and it is here in exceptional cases that researchers may abuse this freedom. This might be a professor who chooses to live far away on the small farm he has inherited and who only comes to the university when he is lecturing or has to attend a meeting, or a research associate who for a period is busy renovating a house and is rarely seen in her office. This can be fine if they take up the situation with the leaders and come to a special arrangement, but a few never do. Some are mistaken in the view that it is within academic freedom (see Sect. 6.6) for them to decide this for themselves. Others may argue that since over time they work much more than they are required to do and no one is directly harmed by their absence, they have the right to take great liberties. In light of what has been said above, it is difficult to look at this as anything other than poor employee morale.

Since the development of the internet and social media, many people feel or believe that it is possible to 'be present' and 'be together' even without physical proximity. However, this 'proximity' is no replacement for being physically present. Text, images and sound on a screen provide only a limited insight into or experience of how other people actually think, feel and behave. The Covid-19 pandemic, which abruptly changed life to such an extent in 2020, gave researchers and research organizations around the world new experiences with this, while demonstrating the importance of physical presence in the workplace. Researchers and research leaders should therefore reflect on what is lost, and possibly gained in addition, when a traditional work environment is replaced or supplemented with a digital one, how the old and new work platforms can complement each other, and how to develop a holistic work environment that can also help promote responsible research practice.

5.6   Serial Conference-Goers and Research Tourism

It is important for researchers to attend conferences, seminars, etc. around the world. Academically, conferences give a good overview of what others are doing and for young researchers they are important learning arenas. Socially, conferences are important for getting to know or maintaining contact with other researchers. They also often lead to concrete research collaborations, and to the exchange and mobility of researchers between research groups. This is especially important for small research communities and research nations.

That said, the attending of conferences and seminars can of course also become an excessive and unwise use of time and money. '*Serial conference-goers*' can be examples of that – scientists (often among the most famous in the field) who travel from conference to conference and only present variations of the same research. The boundaries between what is right and wrong in this respect must be drawn by the individual researcher and research leader. The important thing from a research ethical point of view is that everyone in this assessment is aware that participation in conferences, which fundamentally represents something positive, can also be exaggerated and misused.

Many research groups arrange conferences or the like to present themselves to a national and international audience. Good attendance and the participation of leading researchers in the field are then often criteria for success. The latter contributes to the former. Most of all, of course, the organizer wants the conference to go down in history as an event the research community remembers. In addition to the scientific programme, the choice of venue and framework around the conference is in this respect highly important. Researchers are just like everyone else – they can be enticed and inspired by exciting destinations and experiences. However, if this becomes decisive for participation, it may be relevant to talk about *research tourism*. That is problematic. If, on the other hand, the scientific and collegial programme is decisive for participation, there can be nothing wrong with the framework around the conference being exciting and eventful – the latter can reinforce the former.

In other words, based on their own conscience, researchers here must consider two questions of an ethical nature:

- Where does the boundary lie between responsible and irresponsible use of time and society's resources for participation in conferences and other meeting activities?
- Is participation well scientifically justified, and is the 'tourist experience' not decisive to participation?

Most research organizations have taken responsibility for this and established procedures and practices so that no one is allowed to attend conferences etc. without presenting a paper or poster (unless there are special reasons to do so). It is also common for a leader to have to approve participation. The expected scientific outcome of the participation will then be crucial.

5.7   Minor, Private Tasks in the Workplace and During Working Hours

In practice, all employees perform a number of minor private tasks during working hours and in the workplace. These may be private errands, searching on the internet, updating Facebook pages, private use of the copier, etc. Most employers have a hard time controlling such things, and many also tolerate it as long as it only concerns minor things. The problem occurs when private tasks take over and become a habit, for example, when the copier is repeatedly used to print 3000 flyers for the local sports club or half a day is spent arranging a family holiday. A lesser observed problem is that 5 min here and 5 min there quickly add up to a significant amount of time. Unless the employee is open about this and everything is fully compensated by working in spare time, this should be seen as an abuse of the employer's trust and a typical example of poor employee morale. If the employer becomes aware of this, it should lead to corrective action and, in the worst case, sanctions.

5.8 Work for Others

5.8.1 External Assignments: Paid or Unpaid Work for Others

Many researchers undertake paid or unpaid assignments for others, such as a part-time professorship at a(nother) university, work for publishers, participation in external committees and councils, assignments in radio and television, private consulting assignments for public or private clients, involvement in business enterprises, etc. Some of these assignments can be performed outside the researcher's normal working hours while others need to be carried out during working hours. All research organizations have internal policies and procedures that regulate and limit private external assignments. The main principle is normally that the employee must have an agreement with the research organization prior to undertaking any external assignment. Any exceptions will be specified in the internal regulations. For the research organization it is important to ensure that the external assignment does not:

- Impinge on the main job.
- Burden the research organization's resources or weaken its reputation.
- Lead to a conflict of interest between researcher and research organization.
- Impair the researcher's impartiality and integrity to an undue extent in relation to the exercise of the main job.

Many external assignments can concur with the interests of the research organization. For example, an industrial company will often regard it as an honour if one of their researchers also holds a part-time position at a university, and all research institutions would probably support one of their researchers participating as an expert in preparing an official report for a ministry.

Many external assignments are considered part of the tasks within the scope of the researcher's main position. Examples are work as an examiner, peer reviewer, member of boards and councils in the research community, engagements in professional societies, etc. Many research organizations may have internal procedures that allow researchers to make their own decisions on this type of external assignments, typically when they are one-offs, of limited scope and duration, and have a natural connection to the main job. These tasks are often unpaid or else the remuneration is insignificant. In some cases, however, one must think clearly to distinguish between a private supplementary job and a task that falls within the scope of the main job. An example may illustrate this:

> Example: Due to his clear diction and original style, a professor is quite frequently invited to radio, TV and debates about green energy. The work is done both in and after normal working hours. Sometimes he receives a fee for his contribution, other times not. If the professor's involvement is based on his research-based expertise, this can be seen as the popularization and dissemination of his research to society at large. The university may then view at this activity in two ways: As part of the position as professor, or as a private supplementary job, but one closely related to his main position. In the first alternative the professor may of course spend university time and resources for the assignment and, subject to certain conditions, perhaps in the second alternative too (obviously depending on the internal regulations of the university). However, if the professor is not an expert on green energy, but appears in the media as a committed layman or perhaps even as spokesman for an environmental organization, the activity must be regarded as a purely private activity unrelated to the ordinary job. In this case the university will probably not allow the professor to spend university time and resources on the assignment (again, depending on the internal regulations of the university).

In real life there are many variants of this situation, where the individual researcher and research organization often have to resort to common sense to find solutions if the internal regulations do not provide clear answers.

Most issues related to private external assignments are of a human resources, administrative and legal nature, but there are some ethical issues too, two of which deserve mention:

The first concerns the fact that to varying degrees private work for another organization always impairs the scientist's independence, neutrality and impartiality. How problematic this is depends entirely on the nature and scope of the task and for whom one is working. For example, it is unlikely that the neutrality of a researcher is weakened by carrying out a private consultancy assignment for the Ministry of Education, while a job for an interest organization can more easily associate the researcher with what this organization stands for, thereby weakening neutrality. Similarly, it is more problematic to have a supplementary job for a business that sells goods or services of a controversial nature than for a business that delivers something most people find useful. A researcher who is chairman of the board of an organization will also be more closely associated with what the organization stands for than a researcher who uses his professional skills to help the organization.

The problem that arises when a scientist's independence, neutrality and impartiality is weakened as a result of a private external assignment must be weighed against the benefits the researcher, research organization and society can have of the work. The benefits can then often prove to outweigh the loss of neutrality. An example might be a researcher who, after getting permission from his employer, engages in a start-up company based on the researcher's scientific results. Many will see this as one of the most important contributions a researcher can make – business innovation and the creation of new jobs are essential for society. Many will probably also find it reasonable that the researcher is paid extra for this private supplementary job. Society's support and trust in the researcher, and in research in general, is thus probably strengthened to a far greater extent by such a assignment than it is weakened by the researcher's loss of neutrality in the particular area in question. But not in all cases, because here, as discussed above, the nature of the researcher's assignment and its benefit to society will obviously play a role. Likewise, the researcher's motives will play a role too. In the example here, it is easy to imagine that the researcher's reputation and trust in him can be significantly weakened if the ideal motive of creating a new business obviously plays far less a role for the researcher than personal financial gain.

The most common ethical challenge faced by all researchers when they work for others on a private basis is thus to weigh the positive effects of the work against the disadvantage of weakened independence, neutrality and impartiality. Openness concerning what the assignment involves, why it is being carried out and the researcher's personal gain can prevent speculation and gossip that can give lend weight to critics.

The second ethical question related to external assignments concerns time and attention. The employer will usually demand that the assignment does not significantly affect ordinary work, but in the real world one must expect that this can happen as a natural consequence of time, attention, concentration and energy being allocated to a number of different tasks. If that happens, it may be a violation of the moral obligation of researchers and research organizations to manage society's resources in an effective and responsible manner. However, this must again be weighed against the benefits of the assignment for the researcher, the research organization, the client and society as a whole – an assessment with practical, regulatory and ethical implications.

Research organizations normally have very few opportunities to control and manage how each researcher makes use of their time, attention, concentration and energy on a range of tasks. This puts the researchers' employee morale to the test. If the external assignment is well paid, prestigious, attractive or interesting, it may be tempting to secretly 'steal' time and attention from the main job to a greater extent than that allowed or agreed to with the employer.

5.8.2   Private Engagements Based on Personal Convictions in a Cause

Occasionally, researchers become strongly engaged in a 'cause' either alone or with others in associations, interest groups, protest actions or the like. The starting point is often strong professional or ideological/political views on something occurring in society. An example may be a law professor who engages personally, strongly and extensively in a case in which she believes the judiciary has made mistakes. The rationale may be either academic (for example, the professor may think that the assessment of evidence or the application of the law is incorrect) or ideological/political (the professor may have a special view on how people should be treated by society). Another example could be participation in protests actions against hydro-power developments that will harm nature. Here, too, the motivation for the participation can be both professional or ideological/political or both.

A significant and long-term engagement in cases like this can easily go beyond ordinary work tasks and affect the researcher's reputation, the latter especially if in word and deed the researcher behaves in a way that is inappropriate of researchers. However, some believe that this can be justified when a researcher has knowledge that should be heard, or because it is a civic duty to speak up when one thinks something is very wrong. Because this often also concerns the researcher's freedom of expression, the research organization may find it difficult to exercise its right of governance in such cases, especially if it does not have relevant internal procedures in place.

5.9 Personal Gifts and Hospitality

5.9.1 Gifts and Hospitality in General

From time to time, researchers and research leaders will be offered personal gifts, gratuities, favours, services, or other benefits from collaborators, clients, vendors, etc. The extent and content of this, and the donor's intentions and relations to the recipient, are decisive for whether one should accept or not. This has to do with research ethics because gifts and other benefits offered by others can make it difficult to live up to the ideals of integrity, neutrality and independence. More importantly, both national laws and internal policies and procedures in the research organization will have provisions on personal gifts and hospitality in order to prevent corruption and protect the integrity of the employees and the organization. These explicit regulations are therefore always the basis for assessing what is acceptable in these matters.

The regulations will include all kinds of gifts (tangible and intangible) and hospitality: confectionery, flowers, wine, decorative gifts, gift cards and money, tickets to football matches, theatrical performances and pop concerts, travels, lunches, dinners, hotel accommodations, payment for leisure activities such as golf, hunting and fishing, special discounts on goods, free or discounted professional help, and more.

Although national laws and internal regulations differ somewhat from country to country and organization to organization, the common underlying principle is that gifts and other benefits that can be construed to or intended to affect the recipient's acts of service are prohibited. Gifts and hospitality from clients, interest groups of all kinds, and vendors are then particularly problematic because one cannot disregard the possibility of hidden motives behind the gifts.

On the other hand, gifts and other benefits which are reasonable and proportionate under the circumstances in which they are offered are usually allowed, subject to more detailed stipulations. A few examples may illustrate what may then be acceptable:

Example 1: A researcher is being celebrated on a special occasion, such as a fiftieth anniversary or the last day in a special job or position. There is cake, speeches and presents from colleagues and the head of the department. A few external collaborators and clients from the industry are also there, bringing flowers, a book and a print to hang on the wall. These gifts are of course likely to have a positive effect on the recipient's view of the gift-giver. But since they do not stand out as unreasonable in relation to the circumstances and are given openly as a recognition of good cooperation, it is unlikely that they are given with the intent of influencing the researcher's future actions unduly, or that they might have that effect. Such gifts would then often be acceptable, provided their value is below a limit set by the researcher's organization.

Example 2: A researcher gives a lecture at a seminar organized by an environmental organization. As a thank you, she receives a bouquet of flowers, a bottle of wine, or a decorative glass. Such a gift is not uncommon on an occasion like this one, and it is unlikely that the giver's motive is anything other than to thank the speaker for her efforts and preparations, which will often have taken place in her spare time. Again, such gifts would then often be acceptable, provided their value is below a limit set by the researcher's organization.

Example 3: After a meeting with a client who finances a research project, the project manager is invited to stay longer and have dinner at a restaurant. The client is satisfied with the cooperation and wants to show her appreciation to the researcher. She also hopes the dinner will help both parties to get to know each other better and that this will benefit the cooperation. The intentions of the hostess are transparent and sensible, and after all, one has to eat. Common sense dictates that this mutual influence must be ethically acceptable. It is otherwise good custom to exhibit sobriety on such occasions. If asked about the choice of wine, it's alright to ask for water or a non-alcoholic beverage, or at least suggest something cheap. If asked to choose between a three or six-course menu, the three-course is a natural choice. Otherwise, many organizations have more detailed regulations for situations like this, but it is difficult to foresee everything, so here one often has to exercise discretion.

The characteristics of the gifts and benefits in these examples are that they feel 'right', have little value, are given openly, and do not diverge from what is common for the occasion. Gifts and other benefits that go beyond this may be unacceptable and in the worst case illegal. Here, however, one should be aware that the limits of what is permissible have gradually been sharpened and that some countries and organizations have stricter rules than others. Therefore, the print and decorative glass in the examples above may be in the grey zone depending on the internal regulations of the research organization.

Many countries now take a strict view on gifts and hospitality. In other countries, it is still normal and expected to give gifts and show great hospitality in a professional context, and social interaction is often an integrated part of a professional and business collaboration. To decline gifts and invitations can then be seen as offensive and an expression of not trusting the giver's good intentions. When different and culturally conditioned perceptions of right and wrong clash, a natural solution may be to follow the custom in the host country, as far as this is possible within the regulations at home (national legislation and organizational procedures). The immediate superior should also be informed, if necessary, in order to make a decision.

When all this is said, the caution that one must always keep in mind when receiving gifts and other benefits in a professional context should not diminish the joy of receiving justified attention for meritorious efforts or of experiencing the pleasant sense of having good relationships with others.

5.9.2 Gifts and Hospitality from Vendors

Receiving gifts and hospitality from vendors requires particular due diligence.

Example: A research laboratory plans to buy an instrument worth millions. One of the prospective instrument suppliers invites the head of the laboratory to dinner at a restaurant to discuss the procurement. On the one hand, working lunches and dinners are not uncommon in working life. They are rarely meant as a personal gift or benefit to those who are not paying the bill. On the other hand, one cannot ignore the possibility that in this case the dinner is also an exercise in making the laboratory manager positive towards the instrument supplier. Some suppliers also try to influence laboratory engineers and technicians who will operate the equipment (i.e. not just the decision-maker) with an invitation to try the instrument, perhaps abroad, combined with generous hospitality. In doing so, the supplier may be acting within the law, but from the buyer's point of view, this is problematic. The laboratory must be absolutely sure that employees who have been given responsibility for the purchase only consider the price offers and the technical qualities of the instruments when choosing a supplier. Full transparency on this is also necessary since decisions in a research institution should be open to public scrutiny. The laboratory's internal regulations, therefore, have detailed provisions concerning contact with vendors.

The details of the internal regulations on vendors vary from one organization to another. The principles are usually:

- Just before and during a tender phase, all contacts with potential vendors are limited to strictly business or technical meetings at one of the parties' premises or at another location that is convenient (meeting rooms near the airport or similar). The research organization pays its own costs in connection with visits to or meetings with vendors. The vendors may, however, pay for refreshments and cafeteria-type meals at the vendors' premises. All personal gifts and hospitality offered by vendors are declined or tactfully returned. Exceptions may be minor branding gifts from the vendors, such as a T-shirt or memory stick with the vendor's logo.
- Outside of tender phases, contacts with vendors may have a sober social element, provided it feels natural on the basis of previous and future cooperation and follows the organization's regulations.

Contact with suppliers also requires special care because it is subject to corruption legislation. The example in the box below shows how relatively easy it is to end up as a suspected criminal if one is not careful enough.

> **The Operations Manager Who Had Dinners with a Supplier and Ended Up in Court Suspected of Corruption**
> An operations manager of a Norwegian traffic company was charged with corruption because on three occasions between 2008 and 2011 he had been invited to well-known restaurants by the marketing manager of a long-term supplier (although this has nothing to do with research, researchers can learn from the case). The operations manager ate and drank for a total of NOK 4739 (less than €500), without his superiors being informed. The district court regarded this as 'passive corruption' (on corruption see Sect. 5.11) and fined him. Although the operations manager was acquitted of corruption in the Court of Appeal and by the Supreme Court (September 5, 2014), the case shows how relatively little it takes to be accused of something as serious as corruption. The operations manager had stepped into a grey zone from which it is best to keep a distance.

5.9.3   Invitations to Travel and Events from Clients and Others

Many business companies and public bodies and organizations are interested in utilizing the research communities' expertise, equipment, networks, ideas, students, and more. Some of them work very actively and often over the long term to get in a position to collaborate with selected scientists and research communities, sometimes in competition with each other. The motives vary:

- Some want accessibility to research communities in general or strategically. They can see it as a societal responsibility to support relevant research. They may also see it as a kind of investment that will be beneficial in the long term, for instance when recruiting professionals, by making it easier for their own professionals to follow the research front and the technological developments, etc.
- Some need researchers from outside to perform specific R&D tasks they do not have staffing, time or equipment to do themselves. Some want to use PhD students in their projects. It may take longer, but is often a cheap solution for them.
- Some want to ensure access to qualified personnel by getting universities and colleges to prioritize their fields of interest or having professors formulate master's and PhD theses within these fields.
- Some want to be the first or the only one to exploit the researchers' ideas or results.
- Some have their own solutions, products, or processes they want the research communities to test, use or vouch for.
- Some seek alliances with selected researchers and research organizations in areas where they believe both parties have common interests in political, ideological, technical, and other issues that are debated, planned, or implemented in society.

There is basically nothing wrong with wanting this, but there are several research ethical issues related to the link between the stakeholders' motives and the tools they use to achieve their goals.

One measure, which many use to become better acquainted with those they want to collaborate with, is to invite to events that have both scientific and social content. As discussed above, this can be both straightforward and problematic. This measure is also used by many research institutions when they want to strengthen their position vis-à-vis the business community and the public sector or achieve something special.

An example of invitations that very few will see anything wrong with could be an industrial company that invites researchers or research leaders to visit the company to provide information about its needs and discuss a possible collaboration, while at the same time offering the guests a nice lunch or dinner, perhaps with accommodation if there is a large distance between the parties. Another example could be a company that has established a good and orderly collaboration with a research group and that invites the group to a team-building event with a focus on professional collaboration, but where the team also takes the time to get to know each other better over a dinner. These examples are usually problem-free, partly because the company's intentions are openly stated, the professional work is the main objective, and the social event is expedient to creating a good collaboration (useful for both parties) and cannot be said to be disproportionate.

That cannot be said about invitations to international sports championships, opera performances, salmon fishing, golf, cruises, spas, sumptuous gourmet dinners, etc., especially when combined with travel to an exciting place. It will be extra awkward if the spouses are invited along and the academic content is low or close to absent. Invitations to such events flourished a few decades ago and quickly gained a fairly large scope – it was tempting to taste what it was like to be a VIP and it was flattering to be among the chosen few who received the invitation. Although it is unlikely that a researcher or research leader will provide the host with any ethically unacceptable services in return for this, the threshold is low for an acceptance of such invitations to violate the research organization's internal rules and in severe cases also national laws on gifts or bribes. Some try to get around this by paying the costs of travel, accommodation, etc. out of their own pocket. However, this is still ethically problematic because working hours are spent on events where private temptations have obviously played a role in the decision to participate.

In summary, there are three factors that should be emphasized especially when considering an invitation to trips and events from clients and others:

- That the invitation is natural in light of the relationship with the host, that the host is open and honest about the motives for the invitation, and that one is able to talk openly about it.
- That professional and collaborative matters are in focus and in terms of time constitute the main part of the programme.
- That participation has little effect on one's own research freedom, neutrality, impartiality, or independence.

5.10   Gifts and Hospitality from a Research Organization to External Individuals or Organizations

Research organizations rarely give gifts to anyone other than their own employees, and then almost always to honour an employee on special occasions (50th anniversary, 25 years of service, and the like) according to long traditions in working life. When they give gifts from time to time to someone outside their own ranks, the guidelines for giving are often the same as those for receiving gifts.

On the other hand, it is quite common for research groups and research organizations to invite representatives from other research organizations, the business sector, and the public sector to various types of events with a social element. Their intention is often to present themselves to external stakeholders and key decision-makers to build up goodwill; inform about their research activities; establish or strengthen a strategic partnership; acquire allies in the competition for research equipment funding; position themselves to raise funds for major initiatives and projects, donations, sponsorships, and research contracts; influence the national research policy, etc. At other times, the motive may primarily be to honour and thank external individuals and organizations with whom they have collaborated. The events can range from meetings and seminars to open houses and alumni days; awards and ceremonies; field trips to places of interest, etc., where socializing with congenial hospitality, perhaps also some music or entertainment, often accompanies the scientific programme.

It is obvious that most events of this type are intended to influence the participants in a way that benefits the organizer, and it is important that research communities recognize that this has much in common with similar events that companies and others invite researchers and research leaders to participate in. Due diligence is therefore required, not only as a recipient of hospitality but also as a host. Events should follow accepted traditions and practices, and the hosts should be open and honest about the programme and the intentions behind it. Both guests and hosts can then enjoy both the scientific and the social programme at the event with a clear conscience.

5.11   Corruption and Bribery

Corruption can be defined as 'the abuse of entrusted power for private gain' (Transparency International, n.d.). One form of corruption is bribery, but other acts of dishonesty and abuse are now increasingly defined as corruption. After a number of international organizations began to define corruption as a global problem in the 1990s, corruption has become a criminal offence in most countries, although legislation differs. In 2005 the United Nations Convention against Corruption (UNCAC) also entered into force. It is legally binding and helps to harmonize the views on corruption worldwide and to ensure that nations take steps to fight it. In addition, a number of national and international organizations have been established to monitor the situation in practice. The zero tolerance for corruption that most countries now practice is therefore a relatively recent phenomenon.

In a research ethical context, corruption is unacceptable because it violates principles such as integrity, honesty, fairness, transparency, accountability and legality.

To *bribe* someone is to directly or indirectly promise, offer or give an *undue advantage* to another person (a personal advantage, or an advantage for another person or entity) in order that the receiver shall act or refrain from acting, in breach of his or her duties. The definitions given in the national laws are more elaborate and differ from one country to another. Bribes are usually carried out in secret. The 'advantage' must be significant and can take many forms, both financial (money, gifts, travels, hospitality, entertainments, and more) and non-financial (better grades, honorary positions and awards, and more). However, it is important to note that the term bribery is only used when the advantage is 'undue'. What that might be is defined through case law in each country.

Although corruption is often associated with politics and business, there are many things about research that can potentially tempt corruption. For example, it is conceivable that someone bribes a person in the research community to manipulate research results; provide favourable statements about technology and solutions; manipulate exam results, admission of students and appointment to positions; get a researcher to work on a special field; obtain a contract for the supply of research equipment; receive scientific honours, membership in academies and honorary positions, etc. Researchers and research organizations may also be tempted to *offer* bribes. The examples of corruptive behaviour given here are obviously unacceptable actions from an ethical point of view and will no doubt be breaches of the internal regulations in any research organization (the reader should look this up). Whether they are also criminal offences will depend on the severity of each case and the national legislation.

However, decision-making processes and the transparency that usually prevails in most research organizations provide relatively good protection against corruption. One therefore rarely hears of researchers and research organizations committing corruption or finding themselves in situations where corruption may be an issue. It has occurred, however, and it may of course take place in secret.

However, researchers who collaborate with or operate in countries where corruption is less strictly enforced should be prepared for situations where someone asks for bribes. Some situations may be less serious, such as a public servant who demands money for her own pocket to issue permits necessary to conduct scientific investigations in the country, or a police officer who threatens to arrests a law-abiding researcher unless he receives some money for himself. Acceptance is probably committing corruption. Other situations can be far more serious, such as when a representative of a potential client or buyer of technology and equipment demands money for his own pocket to sign the contract, often disguised in a comprehensive and intricate way as 'consultancy assistance'. A researcher or research leader who is inexperienced and naive when it comes to conditions in corrupt countries, and who is not familiar with international guidelines against corruption and relevant legislation, can make mistakes. Faced with a demand for bribery in such cases, it is therefore always best to immediately stop all activities, present the situation at a high level in the research organization, and possibly consult experts regarding the way forward.

5.12 Network Corruption, Nepotism, Cronyism and the Like

Network corruption, nepotism, and cronyism are terms used when someone by virtue of their own position in a professional assignment or office or the like gives other people undue advantages, often without asking for anything specific in return (if something is required in return, it may be a case of corruption under criminal law). The terms overlap to some extent:

- Network corruption is giving someone in one's own network undue advantage by virtue of one's own position. The term is also used if the whole network is given undue advantage.
- Nepotism is giving relatives and friends undue advantage by virtue of one's own position (as the expression indicates, the term was originally used only when someone in the family was favoured).
- Cronyism is giving 'one's own' undue advantages by virtue of one's own position.

Within research, these forms of abuse of power and trust primarily occur in connection with research management, work in committees, boards and councils, and various forms of peer review. A few examples:

Example 1: A distinguished researcher sits on a committee to evaluate research projects for funding within a public research programme in industrial chemistry. Ideally, it is expected that she puts her own interests aside and in the most factual and objective way possible evaluates the alternative projects based on the program's objectives and requirements, the projects' quality and relevance, and the probability of their success. Without solid reasons, it is then obviously wrong of her to prioritize:

- Her own field of research, analytical chemistry, rather than synthetic chemistry.
- Basic research in the university sector, the sector of the national research system she belongs to, rather than applied research in the institute sector.
- Established and distinguished researchers like herself, rather than newcomers.
- Researchers of her own gender and from her part of the country.
- Researchers she has personal connections to.

Example 2: A researcher sits on a committee to nominate candidates for a vacant research position. Again, the researcher is expected to evaluate the candidates on the basis of competence, merits, and general suitability in relation to the position. So it is wrong if the researcher unfairly allows personal preferences (friendship, gender, where one comes from or has studied, scientific interests, etc.) to influence the assessment.

As with corruption, it is typical of this type of abuse of power and trust that it happens in secret, and often in a cunning way. Rather than arguing directly for one's own interests (which easily raises suspicion), one can, for example, come up with reasonable critical and analytical assessments to evoke weaknesses in the alternatives, and thus indirectly strengthen one's own interests. Network corruption, nepotism or cronyism are therefore often very difficult to detect, let alone prove.

In the examples above, there may obviously be different degrees of bias and injustice. However, network corruption, nepotism, and cronyism are very strong words, the use of which will of course only be justified in the worst cases, i.e. when a person or group falls victim to significant injustice. These are rare events. On the other hand, minor injustices probably happen more often. This form of abuse of power can only be prevented by leaders who are aware of the problem and by effective internal advisory and decision-making procedures, such as the well-developed peer-review systems used by large public research funding organizations, reputable publishers, and others.

This said, all assessments people make are subjective. One must therefore realize and accept that all assessments and decisions made by peers, managers, and others are somewhat coloured by personal preferences and interests – they are never completely objective and fair.

5.13 Supervision of Students and New Employees

5.13.1 Supervision of Master's and PhD Students

All universities and colleges have internal procedures and guidelines for supervision of master's and PhD students that also address research ethical issues (guidance of undergraduates is beyond the scope of this book). Here, as elsewhere, however, it is the practice of such procedures and guidelines that is crucial and which must be given most attention. In this context, four issues are particularly important from a research ethical point of view:

- How the supervisors exercise their authority.
- How the personal relationship between supervisor and student functions.
- How the student is supervised in research ethics.
- How the supervisors utilize the student's work in their own research activities (see also Sect. 16.6.3).

The collaboration between supervisor and student starts when the student expresses interest in taking a master's or PhD in the supervisor's academic field. The research project they eventually agree on must first and foremost be in the best interests of the *student*. Supervisors normally propose a topic for the thesis within their own research interests and plans. This constitutes a potential conflict of interest, but as long as the supervisor is aware that there are two considerations to make, this rarely becomes a problem. In any case, the student must end up with a project that is worth spending time and resources on from a scientific and societal point of view. It should be interesting and challenging, educational, adapted to the student's abilities, feasible within a normal timeframe, meritorious, career-relevant, and career-promoting. It is also part of the supervisor's responsibility to ensure that the project is in accordance with the codes of conduct for research.

Guidelines for supervision of students often emphasize the asymmetric relationship between student and supervisor. The supervisor has authority and power; the student does not, and is often dependent on the supervisor's assistance to complete the work.

Several surveys of the relationship between supervisor and student have shown that most students are satisfied with the guidance they receive, but that some are not. A few find it necessary to change supervisor. Many of them are probably dissatisfied with the scientific guidance they have received, but in some cases, the supervisors severely neglect their responsibilities, abuse their power and authority, or behave badly. Several examples of this have also emerged after whistleblowers received greater protection and students have been able to tell their stories in social media. The examples of experiences are diverse:

- Unpleasant, insulting, degrading, disrespectful, arrogant, or undermining behaviour and utterances from the supervisor.
- Inappropriately demanding, threatening, or controlling behaviour, for example that the supervisor pressures the student to change the thesis work in particular directions that only benefit the supervisor.
- Unreasonable discrimination, injustice or neglect.
- Unpleasant sexual attention or, in the worst case, sexual harassment from a supervisor.
- Supervisors who are absent or unavailable.
- Supervisors who publish results from master's theses as their own work, without giving the students the opportunity to be co-authors (if they are entitled to it), or without referring to them or acknowledging their contributions.
- Supervisors who demand too little of the students or are too present, friendly, indulgent, etc.

Most of these violations of normal good behaviour and responsible research practice have been discussed in general earlier in this book. However, the last bullet point in the list is worth noting. Excessive kindness and amiability can lead to a poor degree of learning. Practice in independent reflection and in giving and receiving objective criticism are key elements of an academic education.

Universities and colleges have schemes to introduce students to research ethics, but the supervisor's daily presence as a role model, coach, and experienced person to talk to about research ethics is even more important. Unfortunately, some supervisors are so focused on their role as scientific supervisors that they neglect this duty. As an example, in several cases where master's or PhD students have been found guilty of scientific misconduct, the wrongdoings could probably have been avoided if the supervisor had been more competent and attentive. These cases have revealed supervisors that:

- Have insufficient competence in research ethics (knowledge, attitudes, and experience).
- Are bad role models, for instance by breaking with responsible research practice, either consciously or through ignorance or carelessness.
- Take it for granted that the students have gained the necessary competence in research ethics through formal courses etc., or think that their job as mentor is done by telling the students to 'read and follow the guidelines for research ethics'.
- Read the draft of the thesis, but do not detect or react to obvious deviations from responsible research practice.

Research ethics often seem to have little place in the consciousness of supervisors who in various ways fail when it comes to mentoring ethics.

5.13.2   Supervision of New Employees

All new employees, and especially those who are entering a research position for the first time, need support and help from the very beginning – administratively, socially, and in terms of research. Many organizations have systematized such support through mentoring schemes and introductory courses. But the time as a newly employed researcher is also an important attitude-forming period when it comes to research ethics and responsible research practice. Few researchers finish their ethical maturing process during their university studies. The extent of ethical learning in any earlier jobs also varies. The mentor's job is to make sure that the new employees know their responsibilities. The collegial environment surrounding the new employees also has a significant influence on this formation process. By being good role models, experienced researchers can inspire the younger ones to work responsibly from the start. The leaders at the lowest level in the organization (including the project managers) are otherwise central when it comes to pushing the ethical formation process in the right direction.

5.14   Whistleblowing in Cases of Suspected Misconduct in One's Own Organization

5.14.1   Responsibility for Reporting Suspected Wrongdoing and Procedure for Reporting

Both national and international guidelines for research ethics indicate the various ways in which researchers who discover or suspect that someone is violating relevant codes of conduct for research are responsible for reporting it to the appropriate authorities. The Singapore statement expresses it this way (World Conferences on Research Integrity, 2010):

> Researchers should report to the appropriate authorities any suspected research misconduct, including fabrication, falsification or plagiarism, and other irresponsible research practices that undermine the trustworthiness of research, such as carelessness, improperly listing authors, failing to report conflicting data, or the use of misleading analytical methods. (Point 11)

Reporting the wrongdoing of other people is never easy. However, few others beyond researchers have practical opportunities and competence to discover research misconduct such as fabrication, falsification, and plagiarism. Reporting such wrongdoing should therefore be seen as an act of loyalty to research colleagues all over. It has to do with the reputation of the research profession.

Employees who report information on certain acts of wrongdoing that they have observed or come across at work have in recent years and under certain conditions been named 'whistleblowers' (employees reporting on something they find wrong or worthy of criticism based on personal and peculiar perceptions of right and wrong, including personal grievances and complaints, are not regarded as whistleblowers). Society has gradually understood that whistleblowers can contribute toward governments, public entities, and private companies operating responsibly, and has established laws to protect whistleblowers against reprisals. However, legislation currently varies greatly from country to country. Among other things, there are differences in the definition of a whistleblower, what they can or should report on, how they can or should report, their rights and duties, and what protection the relevant legislation gives them.

The types of wrongdoing that get the most attention from the public are illegal activities, abuses of power or trust, or activities that can cause serious harm to people or nature. In a research organization this could typically be harassment and discrimination from either employees or management; misuse of public research funds; corruption; experiments or fieldwork that may be hazardous to life, health, or the environment; serious violations of the laws for medical research or handling of personal data, and more. In this book and this chapter, however, *focus is on whistleblowing for research ethical wrongdoing.*

An employee who discovers something that may be a violation of the code of conduct for research should normally report the suspicions to the *employer* in accordance with the organization's internal reporting procedures. The research organization has responsibility for everything that happens in the organization and is in these cases usually the only one formally and practically in a position and with the authority to rectify any wrongdoing and impose personal sanctions on those who may be found guilty. The organization's internal procedures will specify who the reports should be sent to (normally a leader or HR person, perhaps an internal ombudsman for ethics). Reporting via an experienced colleague, employee representative, lawyer, or the like should also be possible.

In cases of violations of laws and public regulations, reporting to relevant external law and regulation enforcement offices may be required in addition to the

internal reporting (both national legislation and the organization's internal procedures may have provisions on this).

In rare cases, a whistleblower may find it necessary to report the wrongdoing in
public through the traditional media, social media, etc. This should be seen as a last
resort if the employer does not handle the internal report correctly and properly, and
only if the case is serious and of public interest. Before going out in public, the
whistleblower must act in good faith regarding the correctness of the suspicion.

The research organizations' procedure for dealing with whistleblowing cases
concerning violations of research ethics norms is discussed in more detail in
Chap. 23.

5.14.2   The Form and Content of the Whistleblowing Report

It is a common belief that the formal threshold for whistleblowing should be low.
The regulations therefore usually set few requirements for the content and form of
a whistleblowing report. However, both the whistleblower and the person or persons
suspected of wrongdoing will benefit from the report being presented in a responsible manner. From a research ethical point of view, the whistleblower should then
act truthfully, objectively, without prejudice, responsibly, and fairly. With such a
mindset, a report regarding research misconduct should be:

- *Justified.* The report must be made in good faith that the matter is worthy of criticism. There must be sufficient facts in the case that indicate that something has
  been done wrong, and the whistleblower must in a conscientious manner assess
  which laws, regulations, or research ethics norms may have been violated. It is
  important to emphasize that one does not want, expect or demand that the whistleblower go deeper into the case than necessary in order to suspect that someone
  has acted irresponsibly or incorrectly. Uncovering the full facts of the case and
  assessing the guilt of wrongdoers are tasks for the research organization's leaders, case officers, and any specially appointed investigators.
- *Presented in a factual way.* The whistleblower should stick to the facts and
  describe what may be matters worthy of criticism in an objective way, in line
  with good practice for case presentations and assessments within research.
  Opinions about the suspect's guilt have no place in a whistleblowing report – that
  is an issue no one can have a justified opinion about without thorough investigations of the case, and after the suspect has also been heard. Personal accusations,
  loose assumptions or speculation (for example about the suspect's motives),
  unfounded and negative statements about the suspect's qualifications, research
  contributions and personal character, etc. do not belong in a responsible whistleblowing report.
- *Presented in a way that contributes to the suspect receiving fair treatment.* All
  whistleblowing reports are the first phase in a case where the actions of a named
  person are investigated internally in the research organization or by an external

body. The principle that everyone suspected of wrongdoing must be treated fairly must also apply to the whistleblower's contribution to the case (see Sect. 23.3). The way the whistleblower issues the report thus matters. Disclosure of the suspicions in the media can, for example, lead to press coverage and prejudice in the research community, which the suspect has little chance of defending against. If the suspicion turns out to be untrue – a regular occurrence – the suspect and others involved may be wrongfully harmed.

5.14.3   Written or Oral Whistleblowing Report

In principle, a whistleblowing report can be made both orally and in writing. The reporting of serious matters, however, entails such a great responsibility that it should be formulated in writing. The research organization may have internal procedures regarding this, which must be followed.

5.14.4   When the Whistleblower Wishes to Remain Anonymous

Most whistleblowers announce who they are when they issue the report. The person responsible for investigating the case will normally in the first instance only make the identity of the suspect and the whistleblower known to those involved in the actual processing of the case (the suspect will then be informed about the whistleblower's identity). Confidentiality during the processing of whistleblower cases is otherwise a rather complicated matter which is described in more detail in Sect. 23.3.

However, a whistleblower should also be allowed to remain anonymous, the reason being that, in principle, one does not need to know who the *whistleblower* is in order to clarify and assess the *suspect's* actions. On the other hand, anonymous reporting is problematic for several reasons:

- It happens that whistleblowing reports are made for malicious purposes – to harm opponents in a conflict, to exact revenge, etc. Information about this is relevant when the case is being investigated. If the whistleblower is anonymous, these aspects of the case may not come to light, and the case will be more difficult for the investigators to process. Many in the research community probably also believe, on the basis of principle, that criticism and accusations directed at others should be made openly.
- Anyone suspected of wrongdoing has the right to know what the accusation concerns, and the right to defend themselves. Defending oneself against an anonymous whistleblower behind the scenes can be additionally stressful.
- Anonymous reporting can lead to speculation that can harm the work environment.

To a certain extent these problems can be avoided if the anonymity of the whistleblower is maintained with the help of respected intermediaries (a respected colleague, employee representative, lawyer, or similar) who can vouch for the whistleblower's seriousness and be a link for communication between the whistleblower and the person in charge of the case. The research organization's internal procedures may have provisions on this, which must be followed.

5.14.5 Abuse of the Right to Act as Whistleblower

As a consequence of whistleblowers now generally enjoying significant public support and protection through laws and regulations, there have also been cases where whistleblowers abuse their rights for their own interests or in order to harm others. An example can illustrate the problem:

> Example: A professorship is announced. The applicants' qualifications are assessed by a committee of three professors from other universities, two of them foreign. Two applicants, both in temporary positions at the department, appear to be the best qualified. One is offered the professorship, the other is not. The loser reacts strongly to this. He goes out to the media as a 'whistleblower' with strong accusations against the institution's leadership. Without giving any factual justification, he claims that one of the peers is biased (she has previously worked in the institute where the competitor obtained her doctorate) and that the head of the institute is practising cronyism (the person who got the position works in the same research group as the head of the institute). He also claims that the management do not like his political views and that the appointment is therefore politically motivated. Due to the form and content of this 'report', the university management initially asks a dean at another university and an external lawyer to assess the application process. Their conclusion is that there is no basis for any of the allegations, that the report is negligent and unreasonable both in content and wording, and not reported correctly. The university's management agrees with this assessment, informs the loser about this, justifies the decision, and encourages him to seek advice from a lawyer or his own employee organization if he wishes to pursue the matter in court. Everyone can feel unfairly treated, and everyone has the right to complain about a decision, but when the main arguments are that leaders have broken laws, regulations, and ethical norms, a complaining researcher is expected to have good reasons for his allegations. It would be normal for the losing party initially to seek advice and assistance from his own employee organization, an experienced colleague, or an employment lawyer and to reason with them about the tenability of his suspicions. Thereafter, he would normally begin reporting internally in accordance with the organization's internal reporting routine. However, the losing party chose another solution: He went out in media, argued unreasonably, demonstrated a lack of knowledge about impartiality, disregarded the factual reasons for prioritizing candidates, and instead emphasized his own speculations about the management's scientific and political interests. By calling himself a whistleblower, he sought to take advantage of the protection that regulations and public opinion give whistleblowers. In reality, this was probably a researcher who, angered about losing, abused his right to oppose a decision and violated several research ethics norms by making untrue and unreasonable claims to the detriment of other researchers and leaders.

Several research organizations have experienced that researchers and others who come into conflict with colleagues or managers can use unacceptable means when their arguments are not heard. Whistleblowing is a new approach to this: While presenting oneself as a conscientious whistleblower, one accuses the counterparty of offences or of research misconduct. It is likely that the despair, resentment, or powerlessness that many may experience in conflicts with others is behind such a course of action.

Despite all that has been said here, whistleblowers must of course always be believed in the first place, and the case should be dealt with on that basis until new facts suggest otherwise. This is as important as a suspect being considered innocent until proven guilty.

5.15   Reporting of Suspected Scientific Misconduct Outside the Workplace

Many breaches of responsible research practice are discovered by researchers, collaborators, clients, and others who are not employed in the research organization where the wrongdoing is taking place. For example, research misconduct such as fabrication, falsification, or plagiarism is often revealed when researchers in the same field or users of the research results around the world discover that something is wrong.

The research ethical responsibility for reporting suspected violations of research ethics norms is the same whether the suspicion concerns someone in one's own organization or elsewhere. Much of what has been said above about whistleblowing will therefore also apply when the suspect works elsewhere. However, those who report research misconduct outside their own research organization will not normally have the protection against retaliation that laws and rules otherwise provide whistleblowers (depending on the regulations in each country). In these cases, whistleblowers should consider discussing their suspicions with managers at their own workplace, and possibly let them report to the relevant authorities in order to lend the report more weight. Based on the principle that reporting should be directed to those with responsibility and authority to process the case, the correct recipients will probably be the suspect's employer and the editor/publisher that has published the work about which there are suspicions. In addition, and if relevant, appropriate national and foreign supervisory authorities should be notified. If suspicious research results have been used industrially or in another way, one must also consider notifying the users, at least if there may be a danger to life and health. The caution and responsibility that is expected when reporting wrongdoing at one's own workplace naturally also apply when reporting wrongdoing elsewhere.

References

Einarsen, S. V., Hoel, H., Zapf, D., & Cooper, C. L. (2020). *Bullying and harassment in the workplace. Theory, research and practice* (3rd ed.). CRC Press.
International Labour Organization. (2019). *ILO Convention No. 190 on the Elimination of Violence and Harassment in the Workplace*. Retrieved September 21, 2021, from https://www.ilo.org/global/topics/violence-harassment/lang%2D%2Den/index.htm

Ipsos. (2019). *Nasjonal rapport: Mobbing og trakassering i universitets- og høyskolesektoren.* [National report: Bullying and harassment in the university and college sector. In Norwegian]. Oslo. Ipsos. Retrieved September 21, 2021, from https://khrono.no/files/2019/08/22/ Nasjonal%20rapport%20-%20Mobbing%20og%20trakassering%20i%20UH-sektoren%20 2019%20(1).pdf.

Statistisk sentralbyrå. (2017). *Utvalgte indikatorer for organisatorisk og psykososialt arbeidsmiljø i Arbeidsmiljø, levekårsundersøkelsen, Psykososialt arbeidsmiljø.* [Selected indicators for organizational and psychosocial work environment in Work environment, living conditions survey, Psychosocial work environment. In Norwegian]. Retrieved September 21, 2021, from   https://www.ssb.no/325268/utvalgte-indikatorer-for-organisatorisk-og-psykososialt-arbeidsmiljo.prosent

Transparency International. (n.d.). *What is corruption?.* Retrieved September 21, 2021 from https://www.transparency.org/en/what-is-corruption

Vranjes, I., Farley, S., & Baillien, E. (2020). Harassment in the digital world. In S. V. Einarsen, H. Hoel, D. Zapf, & C. L. Cooper (Eds.), *Bullying and harassment in the workplace. Theory, research and practice* (3rd ed.). CRC Press.

World Conferences on Research Integrity. (2010). *The Singapore statement of research integrity.* Retrieved September 21, 2021, from https://wcrif.org/statement

Chapter 6
Choice of Research Topic – A Question of Taking Societal Responsibility

6.1 The Authorities' Measures to Ensure That the Research Benefits Society

Researchers and research organizations are responsible to society for everything they do – from the choice of research area and research project, via the research activity, to the reporting, dissemination, and consequences of using the results. The focus on the responsibility of the research communities to society is increasing. Many countries are establishing new strategies and instruments for the interaction between research and society. An example is the European Commission's Responsible Research and Innovation (RRI) policy. RRI represents a more holistic way of thinking when it comes to research, innovation and society; see the box below. The initiative, together with other measures, is a tool for prioritizing in a way that is in the best interests of both society and research, when it comes to spending money on research.

The European Commission's Responsible Research and Innovation (RRI) Concept
Collaboration between research and society is being given increased attention in many countries. In Europe, the European Commission has been central to the thinking about this. In 2012 they launched RRI, a special mindset related to research and a set of tools that has become part of the EU's research policy. RRI became pervasive in the European Commission's research programme 'Horizon 2020' and is increasingly being implemented in national research policies and research programs. The EU justified the measure as follows (European Commission, 2012):

(continued)

© The Author(s), under exclusive license to Springer Nature Switzerland AG 2023
D. Slotfeldt-Ellingsen, *Professional Ethics for Research and Development Activities*, https://doi.org/10.1007/978-3-031-25484-0_6

Responsible Research and Innovation means that societal actors work together during the whole research and innovation process in order to better align both the process and its outcomes, with the values, needs and expectations of European society. (Extract from the lead)

The European Commission's RRI policy covers six key areas:

- *Public engagement*, i.e. involving relevant societal actors in the research and innovation (R&I) processes (in the choice of research tasks, participation in the projects, the use of the results, etc.).
- *Gender equality*, i.e. promoting equality between men and women in research and decision-making.
- *Science education*, i.e. enhancing citizens' knowledge and skills related to R&I and increasing the number of researchers.
- *Ethics*, i.e. focusing on preventing unacceptable research and research practices, and on the ethical acceptability of scientific and technological developments in science and society.
- *Open Access*, i.e. ensuring open access to publicly funded research publications and data.
- *Governance*, i.e. taking responsibility for governing research and R&I for the good of society and the research community.

This can be broken down into a new mindset that the individual researcher and research organization can take as a starting point. It is a matter of working in a way that is characterized by adjectives such as (RRI Tools, n.d.):

Diverse & inclusive: involve early a wide range of actors and publics in R&I practice, deliberation, and decision-making to yield more useful and higher-quality knowledge. This strengthens democracy and broadens sources of expertise, disciplines and perspectives.

Anticipative & reflective: envision impacts and reflect on the underlying assumptions, values, and purposes to better understand how R&I shapes the future. This yields to valuable insights and increases our capacity to act on what we know.

Open & transparent: communicate in a balanced, meaningful way methods, results, conclusions, and implications to enable public scrutiny and dialogue. This benefits the visibility and understanding of R&I.

Responsive & adaptive to change: be able to modify modes of thought and behaviour, overarching organizational structures, in response to changing circumstances, knowledge, and perspectives. This aligns action with the needs expressed by stakeholders and publics. (What is RRI, Process dimensions)

The RRI concept is obviously more important in some subject areas and issues than in others. Some see a contradiction between the RRI concept and 'curiosity-driven' research conducted under academic freedom in academia. However, curiosity-driven research has no less a societal responsibility than other research, but there are alternative ways to fulfil this responsibility.

The societal responsibility of researchers and research organizations is particularly put to the test when choosing research fields and research projects and when using research funds and resources. That is the focus of this chapter.

6.2   The Research Organization's Overall Societal Responsibility when Choosing Research Areas

6.2.1   The Responsibility for Ensuring that the Organization's Research Profile Can Be Defended Scientifically and Societal

All research organizations are established to serve one or more societal purposes – of public or private interest (the term societal is used here and elsewhere in this book in a broad sense). The organization's board and management are responsible for running the organization in a way that is compatible with its purpose. As discussed previously, this is in practice about making decisions on the research profile, what expertise and staffing one needs, which laboratories and equipment to prioritize, how the organization's internal research funds should be used, etc. In the institute sector and in the business sector's R&D units, the management may also decide which projects are to be carried out, although in practice this is most often done by the researchers themselves within the organization's R&D strategy (the researchers' professional and academic freedom to choose topics for research was discussed in more detail in Sect. 6.6).

The board's and management's responsibility for the overall research profile also has a moral dimension, which is not always in focus. One element in this is to secure that the *sum of the organization's research projects and its use of resources reflects the organization's societal responsibility*. Another element is to make decisions about the organization's activities in areas that are controversial in society and among researchers. Examples are research that can be used for military purposes, is aimed at products that can harm nature (but are otherwise useful), or can lead to 'tampering' with life (such as elements of genetic research). Ethical assessments of what is good for society must then be included in the basis for the board's and management's choice of overall research profile. These are difficult questions because the answer can easily be influenced by one's own political, ideological, or religious views.

6.2.2   Responsibility for Changing Research Profile when Society's Needs Change and New Research Areas Develop

The responsibility for choosing a research profile that reflects the organization's societal responsibilities – its mission – is especially put to the test when the importance of entire research areas changes as a result of scientific breakthroughs or major changes in society's needs. Research related to a multicultural society, terrorism, 'green' energy, the internet, etc. were of little interest to society a few years ago but have been given high priority today. New research fields, such as gene technology and computational mathematics, have the potential for a greater scope of scientific discoveries and applications than many 'old' fields. The new is prioritized at the expense of the old. Within language research, today's decision-makers may think that research and education in Chinese and Arabic are more important than in Latin and Greek, and allow this to be reflected in the funding to the research. Such changes in the significance of each research field can have extensive consequences for the individual researcher, as the example in the box below illustrates.

Example of How Changes in Society's need for Research and Education Affect and Challenge Individual Researchers and Research Institutions
Research is conducted to satisfy societal needs, in a broad sense. When the needs change, the content and scope of research also change. When oil was first discovered in Norwegian waters in the North Sea (the Ekofisk field was discovered by an American petroleum company in 1969), Norway suddenly needed to build up an infrastructure to exploit the potential offshore bounty, i.e. to establish new institutions and companies, introduce new laws and regulations, and educate a broad range of professionals. The engineering and technology companies shifted their focus to the new offshore market and built R&D units dedicated to this. New companies were established. Universities started to offer petroleum-related studies and appointed new professors in everything from petroleum geology to petroleum law. The research and technology institutes established new research teams and laboratories for petroleum R&D. Petroleum R&D became 'the great societal mission' for Norwegian research and education for a couple of generations.

As the oil and gas run out or for environmental reasons are no longer exploited, the task has been solved and other tasks will take higher priority. This means that there will eventually be no need for very many researchers who currently work with petroleum. It also means that many petroleum-related laboratories must be looked at with new eyes. The changes will have consequences for the individual researcher. During the transition period, for example, petroleum researchers will notice that the financial support for research decreases or disappears and that students choose other subjects. New research positions and PhD scholarships will not be announced. In the

---

(continued)

institute sector and the industrial laboratories, research positions will be withdrawn and researchers will be moved to other research areas because there is nothing more to do.

In 2014, a university asked for an assessment of whether Norwegian universities' continued involvement in petroleum research and education was ethically sound at a time when the world was trying to switch from fossil to green energy. The request was answered by the National Committee for Research Ethics in Science and Technology (NENT). The committee said, among other things (NENT, 2014, translated from Norwegian by the author):

> NENT finds it striking that the universities do not to a greater extent reflect on their own possible conservative role caused by their collaboration with the petroleum industry. In this connection, the committee would have liked a thematization of overarching questions that deal with the university's response to the knowledge challenges in today's energy and climate realities. What are the key knowledge challenges we face, and how should we meet them? Do the universities as a whole contribute to the further development of today's society, or is it a constructive player in turning the development? What priority should research that helps to prolong the oil age have in the research institutions? (p. 4)

The statement is an example of a research ethics-based expectation that universities continuously inform themselves about and independently reflect on changes in society's needs for research and education, and that they actively restructure their activities accordingly.

The same should reasonably be the case if a research area that was initially a high priority for scientific reasons later becomes less important in relation to new subject areas that have the potential for far greater number of significant scientific breakthroughs.

The responsibility for the organization's adaptation to the large but gradual changes in the significance of different research areas rests, in the first instance, with the *management*. It has the primary tools to make it happen. It can decide which professorships and research positions the institution should have and how internal funds should be used, and is thereby in a position to prioritize fields that are 'hot', and perhaps stifle those that are not. These tools are rooted in the organization's right to govern. It is not the same everywhere. In the business sector's research units and in the research institute sector, the management can normally adapt the staffing to changing needs by moving researchers to other (research) tasks, retraining them, or under certain conditions, dismissing them. This is usually not possible at universities and colleges where professors have tenure or other forms of lifelong job security, often justified by the importance of scholars having academic freedom. In practice, professors sit safely in their chairs until they die or resign, and research what they want within the framework of their position. It can therefore take a long time before a professorship, which has been established, for example, to support the

Norwegian oil and gas adventure, is terminated or transformed into a position in a new, high-priority area. During the transition period, it will therefore mainly be the professors themselves who must take responsibility for their own research turning in a direction so that the position can still be said to be justified, although it may not be as strategically important from a scientific or societal perspective. However, the room for change is significantly narrower for individual researchers than for management. For example, no one can expect that a scientist in petroleum technology switches over to study solar cells, or that a professor in Latin throws himself into Chinese. On the other hand, it may be possible for a specialist in platform construction in the oil sector to start working on wind power construction, and for a professor of Latin to adapt his research to the current realities of the subject. The individual researchers' responsibility for adjustment must therefore reasonably be limited to the professional and financial possibilities they have.

That being said about the individual researchers' and research organizations' responsibility to adapt to changes, it should also be said that:

• In reality, it is often individual researchers and research groups, not management, who take the initiative and lead in the development of new research areas at a research organization, whether the driving forces are changing needs in society or scientific breakthroughs that are becoming important to pursue. The responsibility of research management to continuously adapt the institution's academic profile to the scientifically or socially important tasks is, therefore, best exercised in constructive processes that involve the researchers themselves.
• Major restructuring cannot be realized by the research organizations alone but requires close collaboration between the research organizations and the authorities.
• Universities, colleges, and research institutes also share a responsibility to maintain and further develop a necessary breadth of national competence. This requires good research even in subject areas that are less of a priority.

6.3   The Individual Researcher's Independent Responsibility to Society when Choosing Research Projects

All researchers have an overall *societal responsibility* that must underpin their own research. This societal responsibility is often stated directly or indirectly in national regulations or guidelines for research ethics, or in the research organizations' own policy documents. However, because researchers' societal responsibility is diverse and lies on many levels, there are different opinions about what this responsibility entails. It is also rare that this responsibility is stated explicitly and comprehensively in national and organizational research guidelines. Nevertheless, most people may agree that the core of this responsibility at an overall level can be expressed as follows:

The essence of individual researchers' overreaching societal responsibility is to *actively* do their best to select research tasks and carry out the work so that it:

• Can benefit individuals, groups, or society more broadly (including the research community).
• Does not have unacceptable consequences for anything or anyone.

(This is based on a formulation of societal responsibility used by The Norwegian National Research Ethics Committees (2019, General guidelines, article 12)).

The first bullet point is dealt with further down in this chapter, while the second point is discussed in general in Sect. 6.4 and more specifically in Chaps. 13, 14, and 15.

In connection with the first point, one can ask what the 'benefit' might be. In principle, it will be up to the individual researcher and research leader to provide answers, but the question should also be asked to those concerned, i.e. individuals, groups, and society's various bodies and spokespersons. It is obvious that the answers will then be different, since research can benefit individuals or society in very many ways. Nevertheless, most answers may perhaps be grouped into three main categories. Research that is:

• *Useful for others*, i.e. leading to knowledge, understanding, data, facts, etc. that are needed in public administration and services; in the development of new technologies, processes, products, and services in the business sector; to safeguard the environment, health, and safety; to safeguard nature and administer natural resources; to safeguard and promote art and culture; to direct a critical spotlight on various societal conditions and functions; to help individuals to take informed standpoints, wise decisions, etc.
• *Enriching*, i.e. leading to knowledge and understanding of people, society, history, nature, science, technology, languages, religion, art, etc., which interests, delights and enriches the individual – both laypersons and scholars. This generally requires that the research is followed up with dissemination.
• *Competence building*, i.e. leading to knowledge, skills, etc. that step by step contribute to building up the general knowledge base in various areas, contribute to the development of scientific methods and scientific equipment as a basis for further research, etc. The competence-building effect of research is first and foremost important for researchers within the same research area, but also for the research community and society in general in a broad sense.

Many research projects can be of 'benefit' in more than one of these categories.

Some research projects will be more useful, enriching, and competence-building than others, and there will always be a threshold that must be exceeded before individuals, groups, and the society at large will perceive the project to be to their benefit.

At the same time, many will probably believe that research that is useful, enriching, and competence-building is not only a benefit but is also *important* for society. At any given time, however, there will be differing views on how public funds for research should be used to promote the type of research that may give the best returns.

6.4   Responsible Choice of Research Projects – The Researcher's Most Important Decisions

6.4.1   The Background to the Choice

In relation to many other professionals, researchers have particularly broad freedom to choose their own work tasks. The reason is that the choice of research topics requires deep competence which only researchers in the particular field have. This also concerns academic freedom, which is discussed in more detail in Sect. 6.6.

The choices of research tasks are probably the most important decisions researchers make in their professional career. Take for example a student who knocks on a professor's door to discuss and agree upon a PhD thesis. Two important decisions are then being made: a decision about what the student will spend almost all their time on for the next 3–4 years, and a decision to spend in the order of €300,000 (depending on the country and university).[1] Two everyday decisions, but with major consequences for both the student and society. The responsibility here is thus two-fold: responsibility towards oneself, to spend one's life and abilities on something that is interesting and important; and responsibility towards society to manage its resources in a way that is beneficial to it.

The first major choice of research field and research topic is normally the PhD thesis, which is often suggested or influenced by the supervisor. This choice is especially important because many researchers continue to work in the same field or even with the same or related issues for years afterwards. The choice of topic for the doctoral dissertation is made when the research candidates have minimal competence about what they are choosing. When the PhD is finished, and the career as a researcher is starting, it is therefore important to carefully consider whether the topic dealt with in the thesis is so important that it should be studied further, or whether, from a scientific or societal point of view, it is more important to engage in another research field or topic.

Every choice is based on weighing the advantages and disadvantages of various alternatives against each other on the basis of relevant assessment criteria. On the basis of the societal responsibility one has as a researcher, the choice of the research area and research theme will then basically be to end up with an alternative that is considered scientifically important, which can be enriching for people, or which

---

[1] The total cost of a PhD programme includes the student's costs such as tuition, housing and living costs, and a proportional share of the university's costs not paid by the student, such as the professors' salaries; costs associated with premises, laboratories, and equipment; travel and fieldwork; use of archives, libraries, and collections; administration and office support, etc.

society can clearly benefit from – and which can, therefore, justify society's invest-ment in the research by a good margin.

In all choices, however, other factors also come into play:

- In practice, the options are limited by the *possibilities for funding* the research. Public and private research funds are often directed toward areas where society has a special need for research. This reflects society's own view of and prioritiza-tion of what is of benefit to it, but is nevertheless not the complete answer to societal benefit in a broad sense. This is because money is a limited resource. There will therefore always be important and useful research areas that lack good funding opportunities.
- *Personal ambitions* (desire to discover something interesting, to contribute to creating a good society, developing innovative products, etc.) and *scientific curi-osity* are strong driving forces in many researchers. To succeed in research work, these forces should be unleashed. This favours searching for research areas that are useful, enriching, and scientifically important, and which at the same time fulfil one's own personal ambitions and curiosity.
- A great deal of the R&D carried out around the world takes place in R&D units in the business sector, public organizations, etc. where *the organization's needs govern the choice of research tasks.* The individual researcher's societal respon-sibility is no less in such organizations, but the choice of research tasks will largely be determined by the companies' or organizations' respective business and R&D strategies. In the independent research institutes with a large portfolio of projects on commission for clients – which also account for a large part of the R&D in the world – the choice of research project is correspondingly influenced by the clients' needs and interests.
- *Coincidence or effortlessness* can in practice also direct the choices. For exam-ple, many start up a new project where the previous left off, without asking them-selves critical questions about whether it is really important or useful to go on. Such carelessness, however, is difficult to defend.

Choosing research *projects* that by a good margin are useful, enriching, and/or important for competence building – and steer clear of those that are not – can be easier said than done. One reason for this is that almost all research is a journey into the unknown. Most researchers have, for example, experienced that the research work develops completely differently than planned. Some works cannot be carried out according to plan – some generate completely unexpected results. Many will therefore think that it is difficult or even meaningless to assess the importance of research works *in advance*. However, the fact that there is a risk that the research will yield different results than planned is not an acceptable reason for not choosing projects which, on the basis of one's best professional judgment and given that everything goes according to plan, would seem to be important and useful. A banal example might be a group of archaeologists planning an excavation. In advance, they rarely know for sure what they will find. But they never start digging on pure curiosity alone, and they never dig at random. They do their best to increase the probability of finding something interesting and important by digging where both facts, experience, and professional assessments suggest that they will find something that can justify the costs and use of resources. They also do not start large-scale excavations without first having made less expensive preliminary investigations. The same must apply to all kinds of research. The more knowledgeable and experienced the researchers are, the greater the expectation that they will choose research assignments that justify the use of society's resources needed for the work.

Some believe that it is also difficult to measure or estimate the importance and usefulness of a research work *even after it has been completed*. However, one can always obtain certain indications about this:

- When the project is completed and the findings published, one can always get a certain *indication of the scientific significance* of the results by monitoring how often the work is quoted by others in respected journals. It is difficult to look at many citations as anything other than an expression that the work has scientific significance. Publications that are cited at a more average rate should also be seen as valuable research. On the other hand, publications that are never quoted by others or only quoted sporadically have most likely not produced anything significantly novel, have dealt with insignificant issues or issues of interest only to a limited few, have quality weaknesses, or the like. Some see this as a symptom of what is termed 'wasted research' (see the next section). However, one should be careful to equate very few citations with wasted research. A little-cited published research work, may, for example, have been an important step in the researcher's development within a new field, or it may have provided important knowledge to others *outside the research community* in the public or private sector. Having said that, it is obvious that researchers and research organizations should consider very few citations as a warning that leads to a self-critical evaluation of the scientific importance of the research being conducted.
- Similarly, it is often possible to say something about the *societal benefits* of research simply by asking potential users how useful the R&D results are to them. This is particularly easy to carry out for commissioned research, and some research organizations have their own schemes for receiving feedback from clients on the usefulness of the results. In general, this only gives indications of the *short-term* societal benefits of the research that has been carried out. Some results may prove useful only later, or more indirectly. That is usually more difficult to document.
- Monitoring how enriching and interesting the research is for individuals is also more difficult. Indirectly, however, the scope and reception of knowledge-disseminating measures (popular science articles, books, lectures, TV presentations, etc.) can be a measure of this. This can also be quantified to some extent.

The bullet points here provide examples of how the individual researcher and research leader can get an indication of the importance of their own research, which they can take conscientiously into account when choosing the next project.[2] No more can be expected from the individual. Such self-governance is crucial for researchers striving to make a lasting impression on their fields of research. The individual researcher's will and skills to do this can be stimulated through good research education, a good working environment, and competent leadership in the research organizations.

The fact that many researchers succeed in making good choices of research assignments is proved daily when we, through popular science literature, media, and in other ways, become acquainted with research that will obviously have a decisive impact on societal development, human life and health, and on nature, or research that enchants us with new, interesting knowledge. This cannot only be attributed to chance and luck.

Expectations toward the scope and thoroughness of the evaluations that each individual researcher should make when choosing a research topic must reasonably be limited to something that everyone has the opportunity to carry out in practice. A minimum may then be that before starting a new project all researchers ask themselves a number of concrete, critical questions, which they must then strive to answer honestly and conscientiously. Examples of such both scientific, societal and ethics-related questions might be:

- What do I expect the project to provide in terms of new knowledge, discoveries, data, technology, etc., if everything goes according to plan?
- Why are the anticipated results important?
- Are the results so important that I have no difficulty in justifying the extent of resources used in the project (research time, money, laboratories, etc.) to society?
- Are there indications that others in the research community and society in general (research authorities, potential users in the business and public sector, etc.) assess the importance in the same way?
- Will I be spending a significant part of my life in a justifiable way when I choose this research project over alternative projects?
- Have I thoroughly investigated what others have done in the field before me, and is it likely that the project, if successful, will produce new results that are significant? Or is there enough research in the field so far?
- Are the uncertainties and risks associated with the project justifiable in relation to the importance of the results if it is successful?
- Are the issues in the project formulated in a clear, complete, neutral, and balanced way so that the results of the project will not give a deficient or skewed picture of what I plan to study?

---

[2] The self-evaluation described here must not be confused with the measurements and evaluations of productivity, quality, impact, etc. that research funding bodies and research authorities carry out when awarding project support, evaluating projects and research groups, etc. These organizations use and develop far more ambitious performance criteria and measurement methods, which are often the subject of both debate and research.

- Does the project have a plan to describe and publish both 'negative' and 'positive' results in a neutral and verifiable way?
- How do I ensure that the results are reproducible?

The purpose of this type of question is twofold:

1. To *increase the probability* of achieving useful or scientifically important results.
2. To *reduce the risk* that, in the worst case, a project ends up as wasted research.

This gives no guarantee. Few research projects can, as mentioned above, be planned with certainty about what the results will be.

The objective is therefore not to get unambiguous answers, but that the individual researchers use their professional competence to look for research tasks that stand out as more important than alternative tasks. An example can illustrate this:

> Example: A research group in physical chemistry with ample public funding has purchased a state-of-the-art instrument for determining the composition and structure of substances. Knowledge of the composition, structure, and other properties of substances has been, and is, absolutely crucial both for the understanding of nature and for technological development. A large number of Nobel prizes and other honors have therefore been given to researchers for the development and/or use of such instruments. The group now see it as their responsibility to get the most out of society's investment in the new instrument and their own work capacity. Which classes of substances should they then prioritize, which substances should they start to investigate? Initially there are many alternatives (there are infinite different chemical compounds in nature and others can be synthesized in laboratories), but for them selecting any substance that arouses their curiosity, which they now have the opportunity to study, seems to be a bad strategy – a kind of lottery. They therefore carefully consider a large number of alternatives and end up with a well-founded plan that starts with two substances that stand out as particularly interesting: the first is a substance for which there are already certain indications and theories that suggest that new investigations can clarify certain fundamental chemical properties of the substance and the substance group more generally. The second is a substance where results from other, chemically similar compounds give reason to believe that the substance may have properties with significant technological potential. There are no guarantees what the results will be, but thinking through the alternatives increases the likelihood of obtaining results that will justify the use of society's resources.

Failing to ask these types of questions can only be described as irresponsible. The same must be said about beginning a weakly justified project.

However, these are questions that can be difficult to answer alone, especially for newcomers in a research field. Advice from experienced colleagues, managers and others should therefore be sought. When the research takes place in teams, everyone in the team will be responsible for the assessments. This can help better questions to be asked and answered. If it is difficult to answer such questions, a responsible solution could also be to initiate a preliminary investigation, carry out the work in stages with an assessment of whether to continue after each stage, or otherwise reduce the risk of spending too many resources on research that may be more or less wasted. This, of course, becomes more important the larger the project is.

Some research organizations have procedures for starting a new research project with requirements to assess and justify the project's significance in writing. Where such procedures do not exist, the individual researcher or research group may possibly establish a practice for such documentation. In the next instance, it will generally be relevant to apply for financial support to the project. The research funding organizations will then ask similar questions but in their own way; see Sect. 8.4.

6.4.2   'Wasted Research' – Abuse of Society's Trust

The belief that research is an important tool in the development of society stands strongly in the population. The growth in the number of researchers, research organizations, and research projects from the years after the Second World War to today is proof of that. When society faces major challenges (the shift to green energy, protection against pandemics, migration, etc.), it is therefore not surprising that most politicians see research as a crucial tool in finding solutions. Few, however, are clear about what kind of research they need, and what this research can concretely contribute – in spite of the fact that it must be obvious that only a small part of the total research being carried out will have any direct significance. It is difficult to interpret this as anything other than *politicians trusting* that the research community and research authorities take responsibility for prioritizing the right areas and focusing on the right projects.

While the politicians who govern society seem to a large extent to have confidence that the research system as a whole focuses on important, useful, and reliable projects, there is a certain self-criticism in the *research communities* that not everything is as it should be. This is especially true in medical and biomedical research, where a number of studies have examined the extent to which the results of basic research have led to something that benefits patients. Some of the first findings were startling, such as when a review of 30,000 research papers in biomedicine indicated that, based on certain criteria, 85% of them were actually a waste (Chalmers & Glasziou, 2009); or such as when a study of basic biomedical research in the United States concluded that research to the value of $28 billion dollars (53% of projects) cannot be reproduced (Freedman et al., 2015). However, the figures are very uncertain, the authors state that between 18% and 89% of the works cannot be reproduced, and that the cost of these works amounts to between $10 billion and $50 billion dollars.

At the fourth international conference on Research Integrity in Rio de Janeiro in 2015, '*wasted research*' – as it is called internationally (see box below) – was therefore described by many as the new main problem that the research community must address.

Wasted Research

In recent years the research community, especially in the fields of medicine and biomedicine, has realized that much of the research they carry out is or might be 'wasted'. In a speech after receiving an award from the *British Medical Journal* in 2014, Dr. Iain Chalmers put it this way (Moberly, 2014): 'Medical academia is wasting "massive" amounts of taxpayers' money, and the public must put it under pressure to change.'

Used in this context, the term 'wasted research' might mean:

- Research on issues that are unimportant both scientifically and societal.
- Research in areas that have already been sufficiently illuminated.
- Research projects based on a skewed, unbalanced picture of the problem (bias).
- Research results that are reported in a false, incomplete, or skewed manner.
- Research results that for various reasons are never published (for example, negative results that are deliberately withheld).
- Research results that cannot be reproduced.

A great deal of this is due to poor choice of the research topic and poor planning of the work. It goes without saying, however, that there are nuances in this – some research projects will be more of a waste than others.

Much of this has to do with poor scientific quality, but much is also an expression of poor morals. Wasted research work can therefore largely be avoided if both the individual researcher and the research organization, and those who fund the research, take action to solve the problem.

Those who work with this type of issue are also concerned with finding the reasons why so many scientific works must be characterized as wasted. Many then find an explanation in the researchers' competitive everyday life and the struggle to score points within the systems each country has for measuring individual researchers' performance. Most such systems often focus on the number of publications (cf. Sect. 4.1.2), while the importance and usefulness of the results are measured to a lesser extent (probably because it is more difficult to quantify). If this is correct, wasted research is not only the result of unfortunate circumstances, carelessness, ignorance, or negligence but also of deliberate and irresponsible actions to achieve high scores. In that case, the research community is facing a new ethical problem.

No researcher likes to hear or talk about research being wasted, and the negative reactions range from dismissing it in its entirety to attacking those who raise the criticism. Others see this as an argument that more emphasis must be placed on quality in research, and as a wake-up call to look critically at one's own research as well. Through new studies, some also look for more nuanced and detailed descriptions of the situation that can form a better basis for corrective measures. In any case, the self-criticism that researchers in medicine and biomedicine have shown should act as a role model for researchers in other disciplines as well.

6.5   Caution when Research Can Have both Positive and Negative Consequences

6.5.1   Increased Awareness that Researchers Have an Independent Responsibility for the Consequences of Their Own Research

Individual researchers' responsibility for the consequences of their own research developed into a particularly important ethical issue in the research community when some prominent scientists who had helped develop the knowledge base for making an atomic bomb, and some of those who had been directly involved in developing it during the Second World War (the 'Manhattan Project' 1942–1946), came forward with their concerns about what they had been involved in. Reckless medical experiments performed on humans during the war led to a similar awareness of researchers' responsibilities for the life and health of people in medical research. In recent times, the research community has become aware that many research results form the basis for the development of technology and products that not only benefit society but also harm humans, animals, and nature – and that the responsibility to prevent such use not only lies with the users of the research results but also with the researchers themselves.

In society as a whole, there is a clear trend for people and organizations to take greater responsibility for the *consequences* of what they are involved in – and then not only their own direct actions, but also what they base their activities on, and what follows as a consequence of it. For example, in countries with strict working environment legislation, a company might be criticized for unethical activities if a subcontractor in a country with less strict legislation does not have the same high work standards as the company in question. Likewise, a person buying shares in an equity fund may in someone's eyes be considered an 'accomplice' if this equity fund has ownership interests in companies that conduct unethical activities.

The Limits of How Far the Individual Researcher's Consequential Responsibility Goes

In research, two factors make the assessment of consequences particularly difficult. First, both positive and negative consequences of research are usually the result of *many research projects in a row and in parallel, where many – both researchers and others – contribute towards step-by-step progress over a long period of time.* What responsibility is it then reasonable to place on each individual researcher who contributes something in such long, complex development processes where many factors come into play?

Secondly, there are often both positive and negative aspects to new knowledge and technology. Weapons can be used for defence and attack, medicine has effects and side effects, and information can be both utilized and misused. Should the possibilities for abuse hinder the development of something for the general good?

When are the potentially unacceptable consequences so great that it becomes irresponsible not to take them into account?

An example can illustrate some of the issues here:

Example: A botanist is planning a project to study how a particular plant has developed chemical or biological mechanisms to protect itself in nature. Although the purpose is to understand a part of nature, it is of course also conceivable that the knowledge can be used in practice, for example in medicine or agriculture. Furthermore, it is conceivable that some of these potential applications may also have adverse side effects on humans, animals, or nature, or that the knowledge may be applied in ways that society does not benefit from. What responsibility does the botanist have to assess such potential consequences not related to the sole purpose of the research project? Here, common sense dictates that the botanist's responsibility must be very limited. First, because it is uncertain what the research will reveal, and it is usually meaningless to assess the consequences of something unknown (this is elaborated upon in the next section on the precautionary principle); secondly, because the road from the study of a plant to a chemical or biological product is a very long one. It involves applied chemistry or biochemistry, manufacturing engineering, economics, marketing, and other industrial and commercial expertise far beyond the botanist's professional sphere. Engaging such expertise to carry out a risk, feasibility and impact assessment of unpredictable research results will, in terms of pragmatism, time, and finances, be impossible. The botanist's consequential responsibility must therefore reasonably be limited to the short-term and direct consequences of the project, which the botanist himself – possibly with the support of colleagues and leaders – is able to envision and has the competence to assess. If, on the other hand, the botanist had been actively searching for chemical or biological substances in nature that could be used for special applications or products, the responsibility for assessing and avoiding possible harmful consequences would obviously have been greater.

As indicated in the example, the researchers' responsibility for the consequences of their own research must in practice be limited to performing specific assessments and actions they have the competence and realistic opportunities to carry out themselves. In most cases, it will then reasonably be sufficient for them to:

- Familiarize themself with all relevant laws, regulations and guidelines for research ethics that aim to ensure that humans, animals and nature are not harmed or exposed to unacceptable burdens during or as a direct result of the research work itself. To plan and carry out the research work accordingly. This is discussed in more detail in Chaps. 13, 14, and 15.
- Carry out a conscientious assessment of whether and how the results of the research may harm or burden something or someone, and take this into account in the planning and implementation of the work. In most cases, it should be sufficient for the assessment to be carried out by the researchers themselves, based on their own professional insight, possibly with the support of leaders, seniors and specialists on the type of risk analysis that is relevant in each individual case. Risk assessments with regard to research involving humans are discussed in more detail in Sect. 13.4. In those cases where the risk of harm or burden is high, it will then be relevant to consider whether the project should be carried out or not, or whether special measures should be implemented to reduce the risk. This applies not least in relation to the precautionary principle; see below.

6.5.2 The Precautionary Principle

The precautionary principle was developed at the end of the twentieth century as a guideline for decisions in cases where there are scientifically based possibilities that a particular course of action may cause significant or irreparable harm to nature and the environment, but where the knowledge base is uncertain or deficient. The precautionary principle implies acting so that serious harm does not occur. The principle was quickly adopted in several international treaties and national laws related to nature and the environment, but has also been extended to many other areas. In general, one can say that four conditions must be present for it to be appropriate to employ the precautionary principle as a guideline for action:

- That there is *a scientifically justified possibility* that an act or a failure to act may be harmful.
- That the scientific basis for assuming this is insufficient to quantify the extent and risk of a possible harm.
- That the possible harm is significant or morally unacceptable.
- That there are good reasons to act/not act *now*, rather than wait, investigate or research more. One such reason may be that the possible harm may be irreparable.

There are various formulations of these criteria and of the precautionary principle, and the application of the principle is subject to a considerable degree of discretion, both ethically and scientifically. In practice, the precautionary principle must be weighed against several considerations that are relevant in a decision-making process. In areas where the assessments cannot be based on a reasonably consistent value base in society, people with different political, ideological, and religious beliefs may come to different conclusions when applying the precautionary principle.

The precautionary principle can be seen as part of the principle of caution in research ethics and can be relevant in many different contexts and fields of research (a broad review of the precautionary principle in research ethics can be found in a report by a group appointed by the World Commission on the Ethics of Scientific Knowledge and Technology (COMEST, a UNESCO Council); see COMEST (2005)). In connection to the researchers' general societal responsibilities discussed in this chapter, the principle is especially relevant in two respects:

- Researchers must assess whether their own research projects and their own research results may have consequences that should be assessed on the basis of the precautionary principle.
- Researchers who have research-based indications that the precautionary principle should be considered in a specific case shall notify this in an appropriate way. Climate scientists, who in various ways have 'reported' to the authorities and the public in general about the possibility that human CO_2 emissions can lead to harmful climate change, are examples of this.

The precautionary principle may mean that certain research projects cannot be carried out, or that planning and reporting must be adapted to the risk scenario one is facing. This limits the researchers' academic freedom but at the same time is something one must accept in order to show responsibility towards society. This can be a dilemma if the precautionary principle stops or hinders a research project that can otherwise provide great societal benefits. Research with gene manipulation or the use of stem cells are examples of this. Such research has the potential to provide major medical breakthroughs for the benefit of humans and other living organisms. At the same time, there is an obvious danger that someone may use the knowledge to 'tamper' with humans, animals, and other living organisms in ways many find unacceptable. In these fields, research ethics must be developed step by step in parallel with the scientific development, on the basis of an ethical dialogue within the research community and in society in general.

6.6 Academic Freedom – Also a Responsibility

6.6.1 The Background for Academic Freedom

'*Freedom*' is something everyone values. However, from a historical perspective, freedom has had different meaning at different times and social conditions. The many forms of freedom most people have today – freedom of movement, freedom of choice, freedom of religion, freedom of association, freedom of thought and speech, etc. – have been fought for step by step over generations. Today many of these individual liberties are protected as part of the United Nations International Bill of Human Rights (International Bill of Human Rights, n.d.), which most countries in the world have signed and ratified. Despite this, many people live in countries where by law or in practice they enjoy limited individual freedom.

In large parts of the world, freedom has become an ideal way of life, something that is seen as good in itself – for individuals and for society. The fact that freedom can be seen as a good, however, is entirely conditional on the freedom also having restrictions that prevent an individual or a group from harming other people or society more generally in the exercise of their own freedom, or depriving others of their freedom. The framework for freedom is largely determined by the laws and regulations in society, but to a considerable extent it is up to each individual to draw the boundaries for their own freedom. Freedom is therefore the great challenger of reason and morality.

Academic freedom has its roots in a time back in history were those who studied, thought, wrote, and disseminated knowledge could be subjected to control, persecution, and punishment by those in power. The idea was that academies and universities should be places where, within certain limits, scholars and students could work safely in their search for truth and understanding. Today, academic freedom is used as a collective term for a series of rights society gives to academic personnel and academic institutions. Modern academic freedom is partly based on traditions and scholars' thinking about the fundamental principles of academic activity, and partly on society's belief that the independence and neutrality embedded in academic freedom are essential for academia to function in the best interests of society. Academic freedom is protected by customary law and national law to some extent. In many countries, this protection is strong; in others, less so. Researchers who work in organizations outside academia also have professional freedoms; see later.

Academic freedom takes many forms (see, for example, the United Nations Educational, Scientific and Cultural Organization (UNESCO) (1997, point VI. A. 27)). However, this chapter deals only with the individual freedom that academics have been given to conduct research, and the responsibility towards the society that follows from this freedom.

6.6.2 The Individual Researcher's Academic Freedom in Relation to Their Research Activity

There are somewhat different views on what the individual academic freedom associated with academic research is or should be. However, three elements of this freedom are central and widely accepted:

- Freedom to choose topics for one's own research.
- Freedom to choose methods and procedures.
- Freedom to publish one's own research results.

Here, again, these freedoms cannot be seen as a privilege for the benefit of individuals within a particular profession, but as a tool to promote both good science and good societal development.

These freedoms also have limitations and entail responsibilities. A statement from the European Commission can be taken as an example of this (European Commission, 2005):

> Researchers should focus their research for the good of mankind and for expanding the frontiers of scientific knowledge, while enjoying the freedom of thought and expression, and the freedom to identify methods by which problems are solved, according to recognised ethical principles and practices.
>
> Researchers should, however, recognise the limitations to this freedom that could arise as a result of particular research circumstances (including supervision/guidance/management) or operational constraints, e.g. for budgetary or infrastructural reasons or, especially in the industrial sector, for reasons of intellectual property protection. Such limitations should not, however, contravene recognised ethical principles and practices, to which researchers have to adhere. (From the chapter on Research Freedom)

(The recommendation applies to 'research freedom' and thus includes all types of research, not just research within academia. Some regard this as different from academic freedom).

The three elements of academic freedom listed above present university and college researchers with a number of special challenges and dilemmas. Many of these have a research ethical aspect; some are discussed below.

6.6.3   Limitations on Individual Academic Freedom Related to Research

One of the obvious limitations on academic freedom is that it must be exercised in accordance with research ethical principles. From a research ethics point of view, the freedom to choose a topic for research is therefore not a free pass to conduct research of poor quality, or of low scientific or societal importance. Nor to use public and private research funds and resources irresponsibly or to carry out research that may cause unacceptable harm or burden for anyone or anything. It is rare for someone to consciously use their academic freedom in this way, but anyone who has freedom may in some situations wish to extend it so far that they, perhaps without seeing it themselves, violate other principles that may be more important to follow than the exercise of freedom. Expressed another way, the principles of ethically responsible research must take precedence over the principle of academic freedom.

Another obvious limitation on academic freedom will be the general conditions for research and in particular the opportunities for funding. It is of little help to have the freedom to choose a research topic if one does not have the funds to implement such research. This is an issue that public bodies that determine the economic conditions for university research should obviously take into account. But here, too, academic freedom is not a free pass to obtain funds for one's own research. This is due to two factors. First, one has to take into account *society's freedom* to allocate its funds on the basis of its own priorities. This often results in ample funding of research areas that society finds particularly useful, and then necessarily less funding for other fields. Researchers in the latter areas have less freedom to research in ways they find important. Secondly, there is a long tradition of having to compete with others for project support, and that the project's quality, feasibility, and scientific and societal significance are crucial. Peer reviewers, staff, and decision-makers in the national and international research-funding organizations then decide which projects will receive support. Freedom is thus not absolute – other people's preferences and assessments also come into play.

Having said this, one must nevertheless expect that the authorities, which have given university and college researchers academic freedom, ensure that these institutions are funded and managed in such a way that academic freedom has real meaning.

6.6.4   The Relationship Between the Researcher's Academic Freedom and the Research Institution's Right to Govern

In principle, research activities at universities and colleges are partly determined by the institutions and partly by the individual researchers. The institutions decide which positions they wish to have in different subject areas and the main direction of research for each position. The researchers decide the topics of their research within the framework of their positions. As mentioned above, the institutions can also influence the research profile through the use of the internal funds they have at their disposal. The institutions base their choices on regularly adjusted objectives and strategies for how to realize *their* societal mission. Because funds are limited, management has to prioritize. There are always differences of opinion about the scope of research in different fields and which subjects and projects should be prioritized. Sometimes this leads to acute disagreement or direct conflict, with the management and the researchers standing against each other. An example might illustrate this:

Example: At a university, research on light metals, mainly aluminum, has been prioritized for several decades in order to educate graduates and develop technology that the national industry needs. Prominent professors in the subject have been behind this strategic choice, which the management has supported and seen as an important element in the realization of the university's societal mission.

As part of this strategy, the institution announces a vacancy for a professorship in light metal metallurgy. A qualified physical metallurgist is hired and begins research on formability of aluminum. After a while, however, the new professor receives a random external inquiry about the use of beryllium, an expensive light metal with very special applications in X-ray tubes, instruments in particle physics, space, military, etc. Through this inquiry, the professor becomes so strongly interested in beryllium that she changes to this field of research. Her new research leads to publications in international journals but is otherwise of no interest in relation to the research strategies of the research group, university, or nation.

The research environment around the professor and the institution's leaders react to her choice of research assignments. They do not deny her academic freedom, but regard her choice as violating the prerequisite for the position (the professorship in light metal metallurgy would never have been given to a beryllium researcher). The head of the institute therefore considers 'taking action'. The professor confirms that she has understood the institute's strategy and the expectations of the position, but states that this is not binding on her – her academic freedom gives her the right to choose a research topic within the job description to which she has to relate, i.e. within light metal metallurgy. She also points out that her research results are published in reputable journals.

When the institution's objectives and strategies are in conflict with the interests of the individual researcher, national legislation, institutional regulations, and the employment contract ultimately become decisive. But such cases are also a question of common sense and ethics, and in this case the arguments can then speak both for and against the professor – for example:

Even if the professor had the law on her side, many would probably think that she had abused her academic freedom – that publishing in reputable journals does not make up for her neglect of the responsibilities she has to the research team and for the institution's objectives and strategy. More was expected, and should be expected, of her. She became interested in something else for no other reason than by chance, she broke with the premise of her employment (which she was well acquainted with) and thus made it more difficult for the institution to realize *its* responsibilities to society. The research team of which she was a part was also undermined. An important sector of aluminum research for which the professor had been given the main responsibility was no longer covered. This was disloyal and ethically questionable behavior, some would say.

On the other hand, and even though she had obviously violated the premise of the position, the management of the research institution should perhaps take a step back and consider the case from a different point of view. One of the reasons for protecting academic freedom is that it can lead to research off the beaten track, which sometimes results in new and important scientific insights that no one else has foreseen. Therefore, in this case, it should perhaps also be considered whether the professor's beryllium research is of such an innovative and scientifically important nature that the institution should adjust its research strategy and priorities within light metal research. Again, the details of the case will be crucial here.

This example illustrates the importance of having a good dialogue within the research group and the institution about research priorities and strategies. In this case, it should have been possible to find a solution that could be satisfactory for the professor, her research colleagues, and the institution.

6.6.5 Academic Freedom in Commissioned Research

Research commissioned by a public body, the business sector, and others has gained some scope at universities and colleges. One may ask how this can be defended in the light of academic freedom.

Looking at each commissioned project in isolation, it is difficult to see anything worrying. Academic freedom will be safeguarded by the researchers being able to say yes or no to the assignment. They can also negotiate with the client about the project plan and contract, and can refuse the assignment if no agreement is reached. The fact that both the client and the researchers may have to adapt somewhat in order to meet the other should not be seen as a general threat to independent research – compromises are an ordinary part of any profession.

However, *looking at the totality of commissioned research* at a university or college, one can imagine that academic freedom can be undermined if the activity becomes too great. If commissioned research *initiated by clients* takes over the working life of the individual researcher or research group at a university, the clients' needs and interests can – perhaps without being noticed – become so dominant that the diversity and independence that academic freedom ought to protect can come under pressure. However, quite a lot of commissioned research needs to be conducted before this becomes a real problem.

The tension between the researchers' academic freedom and the clients' interests is not the only and rarely the most important ethical issue encountered in commissioned research. This is discussed in its entirety in Chap. 9.

6.6.6 Various Forms of Professional Freedom for Researchers Outside Academia

Some countries and organizations may have regulations that give researchers outside academia (in the institute sector, the business sector's R&D units, etc.) certain forms of 'academic' freedom – here called professional freedom. However, the researcher's room for manoeuvre is usually narrower than in academia. This is necessary because the research organizations outside academia are established with well-defined purposes and areas of operation. If these organizations are to fulfil their role in society, they must be able to govern research so that the totality of activity fulfils the purpose of the organization. This requires research strategies at the organizational level and the exercise of research leadership, first and foremost when it comes to choosing research fields and strategic projects.

Regardless of this, however, there are arguments that the research institutes and research units in the business sector might benefit from giving their researchers a certain degree of freedom to choose their topic of research. For instance, giving researchers some time for research or studies to dispose of as they wish as a tool to encourage renewal and professional change. Research institutes and research units in the business sector are also dependent on their researchers having a good international network in relevant research fields to stay informed and in a position to acquire new knowledge and technology as it develops. The key to entering such networks is usually to research and publish the results at conferences and in journals. The possibilities for this are strengthened when the researchers allocate part of their time to self-initiated projects. The opportunities for non-academic research organizations to have such schemes will depend on their financial situation. For the institute sector, this will, for example, in practice be conditional on public support and a strategy to profit enough on commissioned projects to finance the arrangement.

Commission-based research institutes commonly practice a form of professional freedom through the way they are organized. Although management is active in defining the organization's scientific profile and can contribute to landing large projects or project opportunities, the main responsibility for obtaining commissioned projects is often placed on the individual researcher or research team. In practice then, the researchers, not the clients, often take the initiative towards commissioned research projects (few researchers sit and wait for clients to come to them with assignments). This gives researchers in the commissioned-based institute sector considerable freedom that contributes to diversity and continuous professional renewal.

References

Chalmers, I., & Glasziou, P. (2009). Avoidable waste in the production and reporting of research evidence. *The Lancet, 374*(9683), 86–89. https://doi.org/10.1016/S0140-6736(09)60329-9

European Commission. (2005). *The European charter for researchers*. European Commission. Retrieved September 21, 2021, from https://euraxess.ec.europa.eu/sites/default/files/am509774cee_en_e4.pdf

European Commission. (2012). *Responsible research and innovation. Europe's ability to respond to societal challenges*. European Union. Retrieved September 21, 2021, from https://op.europa.eu/en/publication-detail/-/publication/bb29bbce-34b9-4da3-b67d-c9f717ce7c58/language-en#

Freedman, L. P., Cockburn, I. M., & Simcoe, T. S. (2015). The economics of reproducibility in preclinical research. *PLoS Biology, 13*, Article e1002165. https://doi.org/10.1371/journal.pbio.1002165

International Bill of Human Rights. (n.d.). *Wikipedia*. Retrieved September 21, 2021, from https://en.wikipedia.org/wiki/International_Bill_of_Human_Rights

Moberly, T. (2014). Waste in medical academia must be addressed, Chalmers urges in the BMJ awards acceptance speech. *BMJ, 2014*(348), g3235. https://doi.org/10.1136/bmj.g3235

RRI Tools. (n.d.). *What is RRI?*. Retrieved September 21, 2021, from https://rri-tools.eu/

The National Committee for Research Ethics in Science and Technology. (2014). *Forskningsetisk vurdering av petroleumsforskning*. [Research ethics assessment of petroleum research. Letter to the University of Bergen 2014 June 18, case no. 2014/3. In Norwegian]. Retrieved September 21, 2021, from https://www.forskningsetikk.no/contentassets/f0cc16a2fb884f838f707ed8b888ade5/forskningsetisk-vurdering-av-petroleumsforskning.pdf

The Norwegian National Research Ethics Committees. (2019). *General guidelines* (English version). Oslo. The Norwegian National Research Ethics Committees. Retrieved September 21, 2021, from https://www.forskningsetikk.no/en/guidelines/general-guidelines/

United Nations Educational, Scientific and Cultural Organization. (1997). *Recommendation concerning the status of higher-education teaching personnel*. Retrieved September 21, 2021 from   http://portal.unesco.org/en/ev.php-URL_ID=13144&URL_DO=DO_TOPIC&URL_SECTION=201.html

World Commission on the Ethics of Scientific Knowledge and Technology. (2005). *The precautionary principle*. UNESCO. Retrieved September 21, 2021, from https://unesdoc.unesco.org/ark:/48223/pf0000139578

Chapter 7
The Ideals of Neutrality, Impartiality and Independence

No one is completely neutral, impartial, or independent in all contexts. Some examples:

- Everyone is dependent on their employer, colleagues, collaborators, funding sources, clients, publishers, etc.
- Everyone has personal interests in their own field of research, which makes them more or less biased when it comes to priorities between fields, use of public funds for research, etc.
- In some contexts, a social scientist who supports or is a member of a political party may not be completely neutral within their research.
- In some contexts, a researcher who performs R&D work on commission for a client is not completely neutral in relation to the client.
- A researcher who takes a stand on a professionally controversial topic for a long time can gradually become party to a professional dispute and develop a self-interest in the case.

In general, there is nothing wrong or immoral about lacking neutrality, being partial, or being dependent on others in relation to an issue. But there are several problems with being too 'close' to objects of research that make neutrality, impartiality, and independence important ideals in research. Two issues should be briefly mentioned here.

One is the danger that, consciously or unconsciously, closeness makes it difficult to be *objective*. Particularly if the closeness affects the ability and willingness to search for, discover, and express facts in an accountable and truthful manner. Objectivity concerns both attitudes and skills. These are developed during research training, and the toolbox of recognized scientific methods and procedures, together with guidelines for research ethics, gives researchers a good foundation to be able to relate objectively to a problem, even though they may have a certain closeness to it. Therefore, people should in principle be able to rely on the research of the social

© The Author(s), under exclusive license to Springer Nature Switzerland AG 2023
D. Slotfeldt-Ellingsen, *Professional Ethics for Research and Development Activities*, https://doi.org/10.1007/978-3-031-25484-0_7

scientist in the example above, even in cases where the scientist's favoured political party has taken a stand.

The second is the danger that closeness to the object of research may arouse the suspicion in others that the researcher is consciously or unconsciously being influenced in an unacceptable way that affects his or her objectivity. Confidence in the researcher and the credibility of the research results may then be weakened. Such suspicions can arise even when they are completely unjustified. In practice, however, there is a threshold for when most people begin to worry about this. Few would think, for example, that a researcher becomes dependent on an external client by performing one assignment, while many will probably think that series of assignments for the same client over a longer period of time can potentially develop both dependence and a positive bias to the client.

However, for most people a researchers' closeness to the subject of their research first and foremost becomes a problem when the research concerns controversial areas or is important for stakeholders and interest groups in society. Even researchers who are in fact completely neutral, impartial, and completely independent, may then experience that others doubt their integrity, perhaps just because the research results may threaten their interests. To reduce the risk of this happening, one must be aware of the problem, strive especially carefully to be objective, and be completely open about all matters that others may think weaken one's objectivity.

Chapter 8
Writing of Project Plans and Applications

8.1 Using Other's Project Ideas: The Risk of Plagiarism

Research is a step-by-step method for gaining new knowledge and increased understanding. Project follows project, often with small thematic and methodological variations. Sometimes, previous works are repeated with new and more accurate equipment, or a new or larger group of informants, etc. It is common practice in the research community that previous project ideas and project plans that have been realized and published can be used further by others, as long as the sources are referred to so that the ideas behind new research projects are well accounted for. This may involve a number of discretionary assessments that from time to time can lead to conflicts between researchers, and in the worst case to accusations of plagiarism. Here are two examples:

Example 1: A professor becomes interested in studying the basic ideological attitudes in a particular category of professionals and what influences their formation. She then builds on ideas that a famous scientist has launched many decades earlier. She publishes her results and duly refers to the older work. A few years later, another researcher gets access to data from a survey on the ideological attitudes of a completely different category of professionals. He then gets the idea to study along the same lines as the professor had done, but now on a more empirical basis related to the new group of professionals where the premises is different. He publishes the results in an article, where he refers to the pioneering work several decades ago, but not to the professor's recent work. The professor reacts to this and states that the researcher's project idea is so close to her own that her work should have been acknowledged and referred to. She describes this as plagiarism of ideas. The researcher defends himself by saying that the basis for the idea for this type of study was laid in the oldest work, that this is internationally recognized, and that there is nothing significant in the professor's work that he specifically uses in his own.

Here, many will probably think that a reference to the professor's work would have been natural. Others will, however, place more emphasis on the fact that the researcher did not present the problem in the work as his own idea, but referred to the first pioneering work, which is more a question of criticizing him for poor referral practice than for plagiarism of

D. Slotfeldt-Ellingsen, *Professional Ethics for Research and Development Activities*, https://doi.org/10.1007/978-3-031-25484-0_8

ideas. However, a closer examination of the details and circumstances of the case would be necessary in order to draw conclusions here.

In the research community, the threshold for what is considered plagiarism of ideas is relatively high. This term is primarily meaningful to use when researchers wrongfully 'steal' a specific and fairly identical project idea from another and present it as their own.

Example 2: A postdoc gets the idea for a research topic which she, together with a professor, further develops into a three-year project. The two jointly write an application for funding to the national Research Council. However, the application does not succeed, and the cooperation ends there. A year later, the professor finds that the project idea can be suitable as a PhD work for a student who has knocked on his office door. He reworks the first project application into an application for a PhD scholarship. The basic idea is more or less the same but the plan is adjusted to the new circumstances. This time the application is approved by the Research Council, and the PhD-student happily begins his work. The postdoc, who has now moved to another university, reacts to this. She claims that the professor has stolen her idea and without her permission used their joint application as a basis for a new project. She regards this as plagiarism of ideas and claims that this behaviour limits her opportunities to pursue her own scientific ideas and interests. As proof, she presents the sketch to the project idea that she had initially sent the professor. The professor, for his part, claims that the project idea is not of such a nature that others could not think the same, that he himself has formulated the scholarship application in his own words, that the plan has been modified, and that they have made no agreement that prevents any of them from using what they previously were together about.

There can be little doubt that the professor behaved unusually and disloyally towards the postdoc by not conferring with her about his new plans and clarifying how the idea from the first project could be utilized in a PhD study. To treat colleagues badly is in itself a deviation from good research practice, but to go from there to accusations of research misconduct by plagiarism is a leap in severity. Distinguishing right from wrong in this case will require an in-depth review of the two researchers' contributions to the first project application. If the postdoc can prove that a central and original element in the PhD project comes from her alone, there may be a plagiarism of ideas. If, on the other hand, and as in many cases, the postdoc has come up with an idea that the professor has refined, where both parties contribute step by step to the final idea, or if the postdoc's sketch was of minimal originality, the wrongdoing might be inappropriate behavior towards a colleague rather than plagiarism of an idea. Again, the details of the case and the circumstances of it will be decisive for the outcome.

The examples show how little it takes to be accused of plagiarism of ideas if the communication and relation with colleagues are poor, and how little is often required to avoid it. In the first example, including one reference would probably have avoided the conflict. In the latter, a phone call from the professor to the postdoc and one extra sentence in the scholarship application would probably have been enough. One extra referral does not matter; one too few can be catastrophic.

The second example also shows that everyone who joins forces to start a new research project must take into account that the project or cooperation may be terminated for completely natural reasons, and establish a collaboration agreement that regulates the parties' rights to ideas and plans at an early stage (see below).

8.2   The Project Plan – Also a Plan for Complying with Research Ethics Norms

Once the topic of the research has been clarified, the project plan is next. In addition to a description of goals and research tasks, the plan also states time and resources needed (staffing, collaborators, experimental equipment, etc.), project organization and management, quality assurance, assumed costs and financing, reporting and publishing, etc. In some research areas (for example medical research) there may be national laws or institutional rules for the design of the project plan. This is not discussed further in this book, which only deals with general matters related to the planning of a research project.

The planning of a research project will always be dominated by scientific and practical issues and choices. However, *research ethical* considerations also have an important place in planning. In the first instance, this happens unconsciously in that the planners' basic ethical attitudes influence their assessments and choices at all times. But in order to take ethics seriously, ethical issues must be dealt with explicitly as well, for example:

- In connection with the planning of the literature reviews of previous works one should:

  - Plan to gain a clear understanding of what others have done in the past, and how the research under planning will contribute something significantly new.
  - Plan to find and review the *originals* of the works one builds on in the new project.
  - Establish routines to ensure accurate references to previous works used in the new research and a balanced discussion of contributions from others.

- In connection with the planning of methods and procedures one should:

  - Choose methods and working procedures that are not harmful to anyone or anything (more on this in Chaps. 13–15).
  - Make sure that the methods and procedures chosen do not give an incomplete, inaccurate or skewed picture of what is being studied (or find ways to minimize the weaknesses and explain them openly).
  - Facilitate for others to be able to control and possibly reproduce the research.

- In connection with the planning of how the actual research work is to be carried out one should:

  - Specify which guidelines for research ethics the work follows.
  - Consider whether there are special research ethics issues related to the work that must be treated with special care.
  - Arrange for all statutory approvals and the like to be obtained, and to establish the necessary routines related to this.
  - Plan to write a daily research log.

- Arrange for ongoing registration and secure storage of research data and research material. Create a data management plan (see Sect. 12.5).
- Create a plan for quality assurance and control which, in addition to covering professional and administrative matters, also contributes to ensuring that research ethics guidelines are complied with (see Sect. 4.2.2).

• In connection with the planning of organizational and administrative matters one should:

- Make sure that everyone (researchers and others) who is to participate in the work is familiar with the research ethics guidelines for the project, including project participants from other organizations and countries.
- Plan measures to create a good collaborative environment within the project in order to promote the participant's performance and prevent conflicts that, based on experience, can lead to unacceptable behaviour (ethical or otherwise).
- Establish routines that ensure that the resources are used in accordance with the project plan and conscientious keeping of timesheets.

• In connection with the planning of reporting, publications, and any use or follow-up of the results one should:

- Make sure that all involved researchers, research organizations, clients, and others understand what each party brings into the project of intellectual property rights (IPR), and agree on the plans for ownership and rights to use the results. Make sure that the participants agree on the criteria for co-authoring reports and articles, how the publication will take place, etc. Conflicts about this are often the starting point for ethically unacceptable behaviour.

In practice, the persons who prepare the project plan should initially sit down and make a checklist of ethical issues that seem particularly relevant in the project. Such a list, and its follow-up, will be good documentation of responsibility.

Next to the choice of research topic, the choice of methods and procedures for the implementation of the research work is the most important element in the planning of a project. The task is to choose the methods and procedures that are best suited for the work and the time and resources available to the project. The right choices are crucial here for a potentially interesting, important, and useful project to succeed. If the wrong choice is made, the project may end up as wasted research. By far the most important here are the scientific methods one chooses. Within all research fields, there is a scientific toolbox with well-documented methods that can be further developed and modified in different ways to adapt to the specific issues in the project. Every researcher in the field is familiar with these methods, and the limitations, accuracy, and uncertainty associated with them. Therefore, the use of these scientific toolboxes is a key element in the argument that research results are particularly reliable (Kaiser, 2019):

'Scientific' is that which competently uses the tools that at any given time are to be found in the scientific toolbox. Thus, the choice of (empirical) method does not have to be decisive as long as a) a method is used at all, b) this method is suitable for solving problems of the given type, and c) the method is used competently.

The entire point of using a method is that it serves as a quality assurance of the knowledge and insights that research produces. Methods should not only provide results, they should also, based on given preconditions, enable the systematic and intersubjective verifiability/quality control of the results. When knowledge is presented as being scientific or research-based, the conditions must be such that peers can review the fundamental data and based on a certain method, assess whether the conclusions are valid. This is the most important prerequisite for scientific quality assurance. (Excerpt from the chapter on the Justification of the norms: methodology)

The choice of methods is also important in a research ethics context. This is because using recognized, well-documented, scientific methods make it easier for others to assess, control, and verify the work so that sloppiness and cheating are more easily detected and thus less tempting to commit.

In the real world, the planning of new research projects easily becomes a routine in which the methodological choices are often made quite unconsciously on the basis of how one usually proceeds, the equipment that is available, etc. This rarely goes wrong, but in order to improve the plan, one should make it a habit to consider alternatives.

8.3   Ethical Issues in Collaborative Projects with Several Participants

8.3.1   Collaboration as a Tool to Promote Responsible Research Practice

Large parts of the research in the world are carried out as collaborative projects between researchers, often coming from several research organizations and countries. Commissioned research is a special form of this which will be discussed later in Chap. 9. Research collaboration can yield great results, and collaboration is therefore often set as a condition for public project support.

In a research ethics context, there are both advantages and disadvantages to collaborating with others:

A positive aspect is the opportunity to be able to consult with others in the project team when faced with ethical issues related to the work. Ethical judgments rarely go completely wrong when they are made by several people. In collaborative projects, the participants also have a certain insight into each other's work. It makes it easier to detect sloppiness and cheating – many reports on scientific misconduct come from colleagues in the suspect's project team (researchers working alone can more easily hide their wrongdoings). Research collaboration should therefore be seen as a tool to reduce the risk of breaches of good research practice.

On the negative side, there are two questions in particular that recur. One is that the parties in the collaboration may have different ethical values and different knowledge and attitudes to issues of research ethics. The second is that for various reasons cooperation with others can also lead to disagreements between the parties. In the worst case, this can lead to conflicts and trigger unethical behaviour.

8.3.2   Caution in Choosing Partners

Anyone who often collaborates with others can be unfortunate and end up with partners one should rather avoid. Some may even get into trouble for a poorly chosen partner's wrongdoings. For example, in a number of the large cases of research misconduct in recent years, where one of the partners in a research collaboration has cheated, collaborators have also come under suspicion, been criticized, or considered complicit.

Trust is the basis for all collaboration in research, but this must not lead to a loss of reasonable degree of caution. In the start-up phase of a research project, when choosing collaborators or responding to an invitation from others to participate in a project, one should therefore also consider the potential partners' suitability in a research ethics context. In practice, this is a matter of exercising a proactive, investigative degree of caution with alarm bells set to ring:

- When something raises the suspicion that the potential partner is unfamiliar with or not so concerned with research ethics guidelines. A conversation about research ethics issues regarding the implementation of the project or a consultation with others who know the potential partner may shed light on this.
- When the potential partner comes from a research organization that does not seem to have a sufficient focus on research ethics and has inadequate internal systems to ensure that good practice is followed. An internet search for the organization or a consultation with others who know the organization may illuminate this.

The fact that warning bells ring does not mean that cooperation should be avoided, only that special measures are necessary to ensure that everyone follows responsible research practice. For example, one can agree on ethical guidelines for the *project*, ensure that everyone is familiar with them, and ensure that the quality assurance of the project captures any deviations from the ethical guidelines (see Sect. 4.2.2). The guidelines in the Montreal Statement (mentioned in Sect. 2.4.3) can also be helpful here.

8.3.3   Special Caution when Collaborating Abroad

International collaboration is an important part of the research:

- Many issues, both scientific and societal, cannot be solved without international collaboration, or they are solved faster and more cost-effectively through international collaboration.
- Many national research communities are small. Collaboration with researchers abroad provides access to complementary expertise, more capacity, special equipment, etc.

- There is a long tradition of researchers applying for positions in other countries, permanently or for periods (many choose doctoral studies, postdoc positions, and sabbaticals abroad). This creates personal relationships across national borders that are often followed up with research collaboration.
- All research is based on research done by others around the world. International project collaboration is a crucial tool to stimulate sharing of knowledge.

However, international research collaborations are associated with some special ethical issues. Four of them are mentioned here:

The Need for a Common Understanding of the Research Ethics Issues in the Collaboration
Although the international research community has a unified view of what is ethically responsible research practice, tradition, political ideology, religion, etc. may cause potential partners to have somewhat different views on right and wrong, both in thinking and practice. The geographical and cultural distance can also be a source of misunderstanding. When entering into collaboration with researchers in other countries, the potential partners should therefore clarify which ethical guidelines they want to follow in the collaboration and how they want to deal with the special ethical issues related to the implementation of the project.

International Collaboration – A Research Strategic Tool
International research collaborations can be rewarding but are also often demanding. They should therefore be rooted in the research strategy of the individual researcher or research group and be well planned. Unfortunately, this is not always the case. Many research collaborations with other countries are initiated because such collaborations are supported by certain public funding schemes (for example, the EU's framework programme for research). Many countries also have bilateral research agreements. And where there is money, there are also researchers. At other times, pure coincidence can also lead to collaboration, for example when two researchers from different countries meet at a congress and, for no well thought-out reasons, end up collaborating. Various forms of opportunism can therefore be said to be behind many international research collaborations. Looking for and following up opportunities for collaboration is generally a necessity in research, but grabbing at everything is unwise. Within the opportunities that exist, decisions regarding collaborations should be based on assessments of how important the collaboration really is, scientifically and socially, and what benefits it can provide that cannot be achieved in an alternative way.

Special Caution when Working or Cooperating with Researchers in Countries with Controversial Regimes
In all countries, research is a political tool. To some extent research is therefore governed by the authorities in the country, primarily through national laws and regulations that can affect research activities in different ways and through the strategic use of public funds for research. Apart from this, most countries want research to be free, i.e. as neutral, impartial, and independent as possible. However, some countries have regimes that want greater control over research. They may, for example:

- Prohibit research that criticizes or challenges the regime.
- Pose inappropriate restrictions on certain types of research or seek to manipulate research results in an unacceptable manner.
- Use research collaboration with other countries to legitimize their own regime (internally or externally) or exploit the research results in a politically inappropriate way.
- Be corrupt, for example by demanding undue compensation for allowing research, funding R&D projects, etc.

This violates internationally recognized research ethics norms. Researchers who plan to operate or collaborate with researchers in such countries must therefore exercise special caution. Dealing with these issues is a difficult task. Therefore, experienced colleagues and leaders should be consulted, and the decisions should be left to the organization's management. The organization may also find it necessary to seek advice from people with local knowledge (for instance local embassies), legal expertise, etc.

Research collaboration in countries with controversial regimes is often *defended on a general basis*, with research being a kind of neutral ground. It is hoped that the contact between the researchers can help to reduce tensions between the countries and that the trust that is created can lead to other collaborations and open channels that make it possible to influence the authorities in the partner country in the right direction. It is also hoped that the contacts at the personal level can inspire researchers in the other country to work for reforms. However, neither national nor international guidelines for research ethics can be understood in any other way than that the individual researcher and research organization must assess the consequences of their collaborations *on a case-by-case basis*. General arguments about the benefits of research and research collaboration do not suffice. One reason is that assessing each project individually often leads to the conclusion that some projects can be defended ethically, some not. For example, it may be justifiable to collaborate with researchers and research organizations in a country with an authoritarian regime if this leads to knowledge and solutions that can specifically benefit people in the country and there is a low risk of harmful consequences. This is acceptable because the usefulness of the results outweighs the risks associated with working in the country.

Research as a Cover up for Espionage, Sabotage and Terrorism
Certain forms of research collaboration can be used illegally to obtain information about national technology and military and societal matters in one's own country. In the worst cases, students and guest researchers can be recruited to conduct espionage, sabotage or terrorism. It goes without saying that this rarely happens, but researchers and research organizations cannot ignore the possibility that it may.

8.3.4   The Project's Guidelines for Research Ethics and the Collaboration Agreement

In almost all research collaborations, the participants' research ethical knowledge and attitudes will vary. Collaborators from different fields of research and different countries may also have different criteria for co-authorship, publication practices, legislation for processing personal data, requirements for HSE in laboratory and field work, etc.

In all collaborative projects, the collaboration agreement between the parties must therefore have provisions concerning a common ethical standard for the project and measures to ensure that all project employees are familiar with the standard, and that they follow it. The extent and content of this point in the project agreement depends on the circumstances, but something can also be said in general:

In the context of research ethics, the agreement between the partners must *as a minimum* have a point that indicates which guidelines for research ethics are to be used as a basis for the work. Example: 'The project will be carried out in accordance with the national code of conduct for research in …'. In addition, if relevant, it should contain additional provisions in areas that often lead to conflicts which may trigger unethical actions by one or more partners. This applies in particular to:

- What rights each party has to the project idea, plan, application for funding, acquired equipment, preliminary results, etc., if the project for some reason does not start or is interrupted, or if one or more parties leave the project before it is completed.
- What should happen if one or more parties does not fulfil their obligations according to the project plan.
- How a possible report of suspicion of breach of responsible research practice within the project is to be handled.
- How the project results are to be published, and what qualifies for co-authorship.
- The project participants' right to access and use other participants' research material.

During the planning of a research collaboration, most researchers are driven to a considerable extent by optimism and enthusiasm for what the partners can achieve together. Few then consider that things can go wrong. However, projects and collaborations that fail are part of everyday life for researchers, often for the most natural of reasons (the project does not receive financial support, key participants become ill, scientific barriers cannot be overcome, etc.). This can largely be thought through and discussed in advance and taken into account in the collaboration agreement.

8.4   Application for Financial Support

Very many research projects cannot be carried out without public or private funding, and writing applications for project support has become a major activity for many researchers. In addition to a reasoned project plan, the funders need specific information about the project's scientific and societal relevance and usefulness (both the direct benefit and the broader impact), plans for quality management, risk control and dealing with ethical issues, the project feasibility, and the applicants' qualifications. The funding organizations' assessments of the applications are largely based on the applicants' own information and statements – the funders' staff and the peer reviewers they ask for advice do not have the time and resources to carry out independent and thorough investigations on this. The requirements for a project application are therefore as strict as for a scientific article in terms of truthfulness, accuracy, objectivity, and openness about the project, the participants' qualifications, the significance of the results, etc.

Because almost all project applications are considered in competition with other projects or other use of money, in practice, they are also *sales documents*. In other words, there is a national and international *market* for funding R&D where researchers compete with each other to sell their projects. This can tempt applicants to suppress risk and other non-promotional factors, and exaggerate the importance and usefulness of the project. This can also happen unconsciously in enthusiasm for one's own research ideas. The result may be an application that gives a skewed description of the project, in breach of ethically responsible research practice. On the other hand, there can be nothing wrong with a project application showing the applicants' enthusiasm for the research they want to start, or that the project's positive aspects are especially emphasized, as long as the proposal as a whole gives an honest, accurate, and balanced account of the project and the participants. Some of the issues that should be given special attention are discussed below.

While many violations of research ethics norms are revealed and made public when the research is reported and published, there are few publicly known cases regarding breaches of responsible research practice in project applications. This may be because researchers are more honest when proposing projects than when implementing them. It may also be that those who ensure the quality of project applications in the research organizations, and those who receive them in the research funding organizations, are not attentive and critical enough, that they do not react, or react outside the public eye. This is certainly also due to the fact that fewer people have access to project applications. The staff at research funding organizations have unique expertise in project applications. Through general guidance and concrete criticism of each project proposal, this competence could be utilized to promote good application practice.

8.4.1 Truthful and Realistic Description of the Project's Goal and Impact

The most important part of a project application is the applicant's description of the purpose of the project and the statement of the project's goal and expected impact.[1] The goal must be relevant to the funding organization. It must stand out as important for scientific, technical, societal or other reasons, and it must be probable that the project provides results that can be utilized in the short or long term. This is often decisive for whether the project receives support. This makes it particularly important that the applicant provides accurate and truthful descriptions and does not exaggerate the outcome of the project. An example can illustrate how accurate the wording should be at this point.

> Example: Through a master's thesis, a research group in physical metallurgy has received indications that a relatively unusual alloying element in aluminum can give the material better forming properties. They want to follow this up with a far more comprehensive series of experiments to see if the effect is large enough to have a commercial interest, primarily for aluminum in certain car parts, where both strength and formability are important. They ask an aluminum company to finance the research on commision and are invited to send the company a project propsal. In the proposal, they formulate the goal of the project as follows: 'The goal of the project is to develop a new aluminum alloy for car parts with better formability than the alloys on the market today'. In the project description, they describe the experimental series in detail. The company agrees to finance the study, and the project is then carried out to the letter according to the work plan. In the final report, the researchers list the results for the various alloy compositions, evaluate their properties when used in car parts, and conclude that one alloy variant shows clearly better forming properties than existing commercial materials. Both the researchers and the industrialists are very satisfied with this outcome. But the result is still miles away from what had been stated as the project goal. No new car alloy has been developed at all, nor has it been attempted. The project was only a start on something that possibly – after many more experiments in the laboratory, industrial testing on a larger scale, and a number of economic and market assessments – might produce a new commercial alloy. A more realistic goal formulation might have been something like: 'The goal of the project is to measure the effect of a selection of alloying elements on the formability of aluminum [the short-term outcome]. Positive improvements in the material properties can be the starting point for later development of commercial alloys for use in car parts [the long-term outcome]'. This formulation gives a more correct and sober impression of the direct outcome of the project, while at the same time making it clear that the work is aimed at an industrially important area.
>
> One may then ask whether those who evaluated the project proposal were misled. In this case, probably not. Aluminum companies have their own expertise on these issues. But a company with less R&D experience might have been misled and expected a result closer to the goal formulation. In any case, it is bad practice to use potentially misleading wording in project proposals and applications.

Most R&D projects are small steps forward in an area that is believed to be important or useful to gain more knowledge about, but there is almost always a long way

---

[1] There are different practices and traditions for the use of terms such as goal, aim, objective, work task, deliverable, effect, outcome, impact, etc. in applications for research support. The readers of this book should 'translate' the terminology used in the text into what is relevant to them.

to go before reaching a breakthrough in understanding or useful results. Realistic information about the outcome of the project in the short and long term is an important part of the truth about the project. However, the outcome one can then hope to reach sometime in the future is almost always more exciting than the direct outcome of each step on the road. It is therefore tempting to emphasize the former over the latter when applying for project funding. Some researchers do this deliberately to make the application more interesting. Others do it unconsciously – many researchers overestimate the importance of their own research.

In recent times, many research funding bodies and research organizations have begun to require that project applicants also account for the expected '*impact*' of the project. The term is not unambiguously defined. Many define it as the short and long-term societal effect of the research (in a very broad sense) beyond the contribution to scientific research. Some also include contributions to the development of scientific knowledge. Some define impact more narrowly as the traditional 'aim' of the project. One may ask whether the latter is not preferable. The most direct effects of a research project are possible to relate to and assess *in advance*, the more indirect and distant ones easily become speculative and uncertain – and few experts have the competence to assess them. *In retrospect*, and especially after some time, it is of course easier to assess the long-term impact of a research project, or at least find examples of it.

8.4.2   *Transparency About Risks in the Project Implementation and Potential Bottlenecks in the Work Flow of a Research Project*

As previously mentioned, it is in the nature of research that many research projects come to a halt or take a different path than originally planned due to unforeseen events. Because this is so common, one must spend time during the planning of the project to consider what could possibly go wrong, what the risks are for that to happen, what the consequences might be, and how risks and adverse consequences can be reduced or avoided. Such issues associated with the feasibility of a project must be given great weight when a project application is being assessed and ranked. It is of little help if the project is important and the applicants are the best in the world if the project cannot, for various, often trivial reasons, be carried through. The absence of this type of assessment is probably also a contributing factor when projects end up as wasted research. An example illustrates how this might come about:

> Example: A group of physicists wants to study the properties of a new type of photovoltaic polymers of potential interest in sensors and solar cells. The material is not commercially available, so the group joins forces with an organic chemist to synthesize the material and a laboratory to make sample cells (a thin layer of the new polymer between two thin layers of conductive material). The production of materials and test cells is estimated to account for 20% of the work in the project. The team applies to the national research council for support, and the application is approved. The chemist launches into the synthesis of the mate-

rial and the laboratory subsequently starts making test cells. Then it turns out that the standard method of making thin layers of photovoltaic material does not work in this case. After some struggling, the team realizes that solving the problem will probably be an R&D project in itself, and in the meantime, the physicists have nothing to do. The project team is stuck. Perhaps this was difficult to predict, perhaps not. In any case, the project is set up with a clear bottleneck in its work flow, the group of physicists could not do anything without test cells, and because it was a completely new material composition, there was obviously a certain risk in the project here. This should have been taken into account in the project plan and also mentioned in the application to the research council.

The ethical issue here is that those who apply for financial support from others must provide information on all matters that are relevant to the assessment of the project. Possible critical elements of work and risks are important factors in this assessment. In a project of great importance and usefulness, one will, for example, accept a higher risk that something may go wrong in the implementation, than in a project of medium importance.

Failure to *assess* the risk that something may not go according to plan is therefore negligent, and failure to disclose *known risk factors* is a violation of the norm of truthfulness in research ethics. Both are reprehensible. Honest risk assessments in a project application are a plus; the absence of risk assessments should arouse suspicion.

8.4.3   *Truthful Description of the Project Participants' Qualifications and Merits*

All project applications include a description of the project participants and the research organizations where they work, with CVs attached. The application templates will have requirements for what information must be provided. Those who evaluate the application must first ensure that the applicants are qualified for the task and able to carry out the work according to plan. However, project applications are often assessed in competition with other applications and are then also ranked according to who is *most* qualified and equipped. Applicants, therefore, have good reasons to present themselves in the best possible way. Basically, there is nothing wrong with that, as long as the CVs and the statement of one's own qualifications and resources are not misleading or untrue, or otherwise violate good practice. Serious falsifications are probably extremely rare, but embellishing the qualifications in a misleading way can tempt some. Two examples can illustrate this:

Example 1: A researcher applies for support for a project in a subject area where she has on two occasions previously supervised master's students. Instead of writing that she 'has supervised two master's students in the field', she states in the application that she 'has worked in the field for the past five years'. This is not a direct lie, but most funders will probably perceive the choice of words so that the researcher has worked far more extensively and personally in the field than she has actually done. The wording is thus misleading and a violation of research ethics norms.

Example 2: Another researcher applies for support for a project where experience in industrial business development will be an advantage. The applicant's industrial experience is very limited, but he has on some occasions given a local company some technical advice on production. Instead of reporting this as it is, he writes in the application that he 'has experience from consulting assignments in the industry'. Again, the statement per se is correct, but the choice of words leads the funders to believe that the applicant has broader and more relevant experience. Here, too, the formulation is therefore misleading.

The degree of deception in these examples is moderate, but the main point is that in all project applications one must strive to bring out the positive in the application without using formulations that can be misleading. The best thing is to say things as they really are.

What has been said here about the description of the project participants' qualifications, of course, also applies to the description of the research organizations involved.

8.4.4   The Requirement that Applicants for Project Support Account for Research Ethical Issues

The national research councils, the European Commission, and other major research funding organizations demand that the projects they support must follow responsible research practice, and can explicitly state which guidelines for research ethics one must follow. They also require that everyone who applies for project support according to more detailed guidelines must account for relevant ethical issues in the project. Research ethics are included in the assessment criteria for obtaining support, and breaches of recognized research ethics norms must be reported as deviations from the project agreement.

8.5   Caution with Whom One Gets Project Support from

When researchers and research organizations receive external, financial support for their research projects a 'relationship' between the recipient and the funder (sponsor, donor, client) always develops. The consequence of this is that the choice of funding source(s) can have an impact on the credibility of the project results, since researchers and research organizations are often associated with the interests and reputation of the organizations that fund their research. An example can illustrate this:

Example: A research group wants to compare the living conditions of a microorganism on the seabed around and far away from an ocean fish farm. In addition to support from a national research council, they seek support from other sources, partly to increase the scope of the project and partly to get input on their research. Suppose they have four options:

1. Support from one or more environmental organizations.
2. Support from a national aquaculture industry association.
3. Support from one or more coastal municipalities.
4. Support from both environmental organizations and the aquaculture industry.

Although the research is carried out completely independently of the sources of funding, and the financial support is in the form of donations (i.e. no form of commissioned research), it is obvious that many may perceive the research in different ways, depending on what the source of funding is. In the first two alternatives, there is an inherent possibility that some will worry that the research may be affected by the respective interests of the environmental organizations or the aquaculture industry, in a way that may weaken their confidence in the results. Such an inherent possibility is also present in the third alternative, but probably to a much lesser extent. Many will then also be unsure of which direction the influence may go (a municipality has many interests). In the fourth alternative, the opposite can happen – that the sources of funding help to strengthen the credibility of the results. When parties who often have opposing interests come together to support research, it is conceivable that most people will feel a sense of assurance that the researchers are neutral.

Reflections on these and similar issues related to potential sources of funding are primarily relevant when the project addresses politically or ideologically controversial issues, when the sources of funding are unusual or unfamiliar, and in research on commission from clients.

Reference

Kaiser, M. (2019). *Research values*. Oslo. Retrieved September 21, 2021, from https://www. forskningsetikk.no/en/resources/the-research-ethics-library/systhematic-and-historical-perspectives/research-values/

Chapter 9
Commissioned Research and Other Assignments for External Clients

9.1 The Motivation for Commissioned R&D

Many research organizations carry out assignments for public and private customers. The assignments are based on the researchers' scientific skills, ideas, ability to innovate, professional networks, professional ethics, the organizations' facilities, IPR, ability to manage commissioned research, neutrality, reputation, and more. This especially applies to research and technology institutes that are founded primarily to serve the need for R&D and related services in various sectors of society, and where the activity is characterized by commissioned projects. Commissioned R&D is associated with a number of distinctive research ethical issues.

Many researchers are particularly interested in applied research and development. For them, the opportunity to contribute directly to the development of society and see the results of their own work within public administration, business, nature and the environment, cultural life, etc. are the great motivation. This requires closeness to societal issues and collaboration with the potential users of the research results, preferably from beginning to end. This also opens up greater opportunities for users to contribute to the funding of the research. Therefore, commissioned research often arises in the interaction between researchers oriented towards applied research, who have an overview of the research front and special research resources, and users with special needs and interests. Many clients also have managers and professionals that are highly qualified to collaborate with academia and that have often received the same or similar education as the researchers, perhaps even at the same institutions. Some are researchers themselves and adhere to the guidelines for research ethics like all other researchers. Technology oriented companies, for example, often have their own R&D units that also outsource assignments to others. When this is the case, the assignments often take on the character of a collaborative project between colleagues of profession. The fact that one party is paying for the collaboration, and that the parties have different tasks and rights, constitutes an

© The Author(s), under exclusive license to Springer Nature Switzerland AG 2023
D. Slotfeldt-Ellingsen, *Professional Ethics for Research and Development Activities*, https://doi.org/10.1007/978-3-031-25484-0_9

inequality that affects some aspects of the partnership but which is often irrelevant in the day-to-day collaborative work.

Many research organizations that carry out commissioned R&D also undertake other knowledge-based services, such as consulting services, laboratory services, etc. on the basis of the resources and the competence accumulated through research. These assignments represent an opportunity to give something *directly* back to society, which after all has funded the researchers' knowledge building. Such assignments also provide contact with the real world that can be instructive for researchers.

Researchers and research organizations are always responsible for ensuring that the commissioned work is carried out in accordance with relevant guidelines for research ethics. Some of the issues that should be given special attention in commissioned research are discussed in Sects. 9.4–9.6. These sections also include a number of issues that researchers from time to time may perceive as problematic in relation to clients. For the sake of balance, it should also be noted here that from time to time clients may have problems with researchers, but this is mentioned in the following only to a minor degree. When focusing on problems, it is important not to lose sight of the broader picture: Every year, thousands of commissioned R&D projects are carried out in which the collaboration between researchers and clients runs smoothly.

As an introduction to this, however, it is appropriate to say something about commissioned R&D more generally. This is done in Sects. 9.2 and 9.3 and in the first part of Sect. 9.4.

9.2 A Definition of Commissioned Assignments

There are many different traditions and forms of commissioned assignments. That has led to different perceptions of how commissioned R&D should be defined. In a book such as this, a broad definition that covers several types of commissioned work is appropriate. One such definition may be:

A commissioned assignment is R&D or another service that is performed to satisfy identified needs or interests of an external client. The assignment is carried out in the form of a well-defined project which is principally financed by the client. The topic, the goal, and the scope of the work are agreed upon between the client and the research organization or the researchers who are undertaking the assignment. Ideally, the agreement between the parties regulates all matters related to the work and its results.

'Commissioned research' is often used as an abbreviated term for research, development, and other expert assignments carried out in a research organization, even when the pure research component of the work is modest. It is important to distinguish between research *on commission* for others and research *with financial support* from others (sponsored research). One can say, somewhat imprecisely, that the former are projects that 'belong' to the client, while the latter are projects that 'belong' to the researchers or the research organization. A person who donates money to research is thus normally not to be regarded as a client for the projects financed by the donation, even though there may be conditions attached to the donation. Public or private sponsors that support research projects are not clients either, even though there are always conditions attached to their support.

Some projects are based on collaboration between researchers, users, and funding sources where all parties have an interest in the project in one way or another and may contribute to the implementation in different ways. In such cases, it will be the provisions of the cooperation agreement between the parties that determine whether the project is commissioned research or not.

The broad definition of a commissioned assignment used here encompasses many variants of collaboration. For example:

- One or more clients and one or more research partners.
- Several ways of financing the projects.

    - The most common is that the client pays the costs of the research, sometimes also with some form of public support.
    - Sometimes the research organization also contributes to the financing of the project, for example by not charging the client for project ideas, use of previous research results (background IPR), software and special equipment used in the research, or for certain work tasks. The condition is then usually that the research organization receives something in return in the form of royalties from sales revenues, or the right to use or own the results or parts of the results, etc.

- Many forms of work-sharing between the client and the researchers.

    - In some projects, there are reasons for the researchers to work completely independently; in others, that the client contributes to the project work in different ways. Occasionally it may also be necessary for client and researchers to work closely together during the implementation of the project.

- Many ways of arranging the ownership of the results and the rights to use the results.

Within the definition of a commissioned assignment given above, some would argue that the 'external' client could also be someone from the same organization (for example, the organization's management or another unit in the organization). Assignments for someone within the organization can be organized, managed, and administered as ordinary, external commissioned assignments, but in important contexts, there is a significant difference between a real internal and external client. It is therefore natural to distinguish between them, for example by calling one 'internal assignments' or 'internal research' and the other 'commissioned assignments' or 'commissioned research'.

9.3   How Commissioned Projects Are Initiated

Commissioned projects are mainly initiated in two ways:

- By *the researchers contacting a potential client* with a project idea or a project proposal that they believe the client will benefit from or be interested in, and which they offer to carry out on commission for the client. The starting point is then usually an original project idea; original know-how, results, solutions, patents or other intellectual property; special equipment, and more. This approach is a necessity in institutes with low basic funding where most of the institute's revenue depends on project initiatives taken by the researchers. If the potential client finds the proposal interesting and has the funds to finance the project, the next step will be to discuss the details regarding content, scope, implementation, etc., and to prepare an assignment agreement.
- By *a client contacting one or more research organizations* for help in carrying out a specific R&D task. Many clients, first and foremost in the public sector, must then follow statutory or regulatory procedures for the procurement of services of a certain scope, primarily in order for there to be open competition for the assignment. The most common procedure is then for the client to announce an invitation to tender where the assignment and the award criteria are further defined. Those who wish to be awarded the contract must then present their solutions and conditions in the tender, and neither party can normally negotiate the tender after it has been submitted. In the case of smaller projects and in companies and organizations that are not subject to requirements for competitive tendering of services, the client often turns directly to researchers or research organizations they assume to have the best competence and resources, or whom they know and trust. This is the best way to develop such a project because the knowledge and experience of both the researchers and the users can then be fully utilized in the planning. The use of tenders can, however, be an appropriate procedure even when it is not required by law or regulations, especially if it is important to prove the neutrality and independence of the researchers. An example could be when a company wants to engage a research group to review elements of a company's business or products, to be used as documentation to public authorities.

In both cases, one must carefully consider whether there are scientific, ethical, or other reasons for *not* entering into cooperation with a potential client. Assignments for suspicious clients or clients involved in illegal or ethically unacceptable activities must be declined.

9.4 The Assignment Agreement

The assignment agreement should ideally attend to the interests of both the client and the researchers or research organization (the 'contractor'), and each party is responsible for promoting its prerequisites for entering into the cooperation. Without an agreement there can be no assignment. In order to reach an agreement, all parties must show understanding of the other's needs and interests related to ethics, copyright, finance, ownership and access rights, etc. In the discussion about this, the researchers and research organizations have the primary responsibility for ensuring that no provisions in the agreement prevent the work from being carried out in accordance with relevant guidelines for research ethics, unless there are special circumstances that make it ethically acceptable to deviate somewhat from the guidelines. Without agreement on that, the researchers must decline the assignment.

Many of the ethical issues associated with commissioned assignments are rooted in the project agreement. They arise especially when the contract is negotiated and during the implementation of the project, as a consequence of ambiguities or shortcomings in the agreement. From experience, there are three elements in the assignment agreement where the parties must in particular be on alert. They are discussed in the sections below.

9.4.1 The Provisions on the Project's Content, Goals, Procedures, and Methods

Everyone who plans a research project must reflect on ethical issues related to the research topic, as discussed in further detail in Sect. 6.3–6.5. This naturally also applies when researchers submit a tender or negotiate with a potential client about a commissioned assignment. However, the choice of topic seldom stops an assignment for ethical reasons. When it comes to the content and scope of the project, and the choice of methods and scheme of implementation, the ethical issues may be somewhat more complicated. Here one must distinguish between projects that are up for tender and projects that are not:

When the Project Is *Not* Up for Tender
In this case, the parties can cooperate on the planning of the project. Most clients let the researchers prepare a draft project plan but prefer to come up with a draft assignment agreement themselves. Many clients have personnel with extensive experience related to the issues to be studied. Some also have their own research expertise. Their views can then help to make the plan even better than the researchers can alone.

When the parties can work together on the planning, they usually find ways to carry out the project that satisfy the professional and ethical standards of all involved. Exceptionally, however, one can meet a client who for one reason or another insists on using procedures and methods that in the researchers' view may be in violation of their professional ethics by giving an incorrect, skewed, or incomplete picture of the issues that the assignment is intended to clarify. Sometimes, this is a good reason to *decline the assignment*. Other times, it can be justified to take on the assignment under certain conditions, for example if the project, despite its weaknesses, may generate some useful and justifiable results. In this case, the assignment agreement must in addition allow the researchers to be open about what they believe are limitations and methodological weaknesses in the project and about the consequences for the accuracy, completeness, and reliability of the results. This must then be expressed in writing to the client before the project starts and in all documents describing the project and the results. Those who read the project reports and consider using the results will then be informed about the shortcomings and uncertainties in the work and the result's limited value as a basis for possible decision-making – as the researchers see it.

When the Project Is Up for Tender
In tender competitions, the research topic and project goals are defined by the client in the invitation to tender. In addition, work tasks and procedures may be outlined, and some specify an upper-cost limit for the work. Sometimes the researchers who consider bidding for the assignment, have objections to the tender invitation. One example is objections to the scope of the project. As when a hydropower developer wants a research organization to carry out a statutory impact assessment, but the invitation to tender limits the work to include relatively 'harmless' factors in the hope that the licensing authorities do not react to it. Another example is objections to the detailing of procedures and methods, as when for financial reasons a public agency limits the number of people who are to participate in an interview survey so severely that the result, as the researchers see it, may give too uncertain and skewed a picture of the issues to be studied. A further example is objections to the cost limit set by the client, which the researchers may consider too low to solve the assignment in a professionally sound manner. The task for the research organizations that, despite this, choose to submit tenders, will be to offer a professionally and ethically responsible solution as much as possible within the framework set by the tender conditions. This should be done in a factual and professionally well-founded way, and experienced colleagues and leaders should assure the quality of the assessments. The client is thus given the opportunity to think about and possibly adjust the setup, and the researchers get a documentation of their responsibility, openness, and neutrality towards the client. Good clients will look at this as an expression of competence and responsibility; bad clients may dislike it and award the assignment to others. If the tender is accepted, the information about the methodological weaknesses and their consequences must of course also be stated in the project reports.

9.4.2 Provisions on Ownership and Licence Rights

The results of a commissioned research project are partly intangible (the intellectual content of the reports, general knowledge and know-how gained in the project and the like), partly tangible (the physical reports, equipment, materials, samples, prototypes, etc. that have been purchased or developed). The client and the researchers must agree on what rights each of the parties shall have to these results, both those that are planned and those that for unforeseen reasons may come in addition. In the first instance, a distinction is made between ownership rights (usually the right to use the results, make money from them, sell or transfer them to others or dispose of them in another way, such as patenting them or keeping them secret) and licensing rights (usually the right to use the results for specified purposes and on specified conditions). The scope and content of these rights can be agreed upon in many ways, including various arrangements for shared ownership.

Basically, this is an everyday negotiation between a buyer and a seller of services, where one has to find solutions that both parties can accept before an agreement can be signed. The outcome of the negotiations depends on the circumstances of the assignment, but the following is quite common:

When the client is a *business enterprise (private or public) and pays all the costs of the project*, the starting point is usually that the ownership of the results goes to the client. One usually gets a reasonable right to dispose of whatever one has paid for. But several deviations are common:

- The research organization often gets the right to dispose of the results with regard to applications outside the client's commercial area of interest (this can be arranged in several ways, through shared ownership or various license arrangements).

  Example: A mathematician is commissioned by a large chemical group to improve an algorithm in a simulation programme that the group has developed for one of its chemical processes. However, the mathematician envisages that the algorithm can also be used in other contexts, and negotiates that the research organization will have the ownership of the assignment results, while the chemical company will have a royalty-free, exclusive right of use within its own business area. This satisfies the company's needs and prevents its competitors from gaining access to something the company has actually paid for. At the same time, the company waives the rights to a possible commercial sale of software for other purposes, based on the improved algorithm. The company accepts this gladly since few major enterprises spend resources on making minor profits (which would be the case here) on activities outside their core business. The fact that the research organization in this case received the ownership rights and the client a limited licence to use the result was simply because the parties found this arrangement easier to formulate and practice than the opposite alternative.

- The research organization often gets exclusive rights to all *unforeseen* results, which in practice the client has not paid for (this can also be arranged in several ways, through shared ownership or various license arrangements).

  Example: A research group in electronics is commissioned by an industrial company to develop a tailor-made gas sensor to detect changes in a special process parameter during the manufacture of one of the company's products. It is agreed that the company will have the ownership of the tailor-made sensor, while the research organization will have the ownership of all other results of the assignment. The first part of the work is to try out different detection principles. A standard method stands out as the best for the purpose, and the researchers, therefore, construct the sensor on the basis of this principle. Four units are then produced, installed, and successfully tested in the company's factory. The assignment has thus been completed, and the client is very satisfied. However, during the work, the researchers had also come up with and tested an unortodox detection method. It turned out to be completely unsuitable for the client's use, but the experiments indicated that it could have significant potential in environmental monitoring. This was an unforeseen result, possibly of commercial value, to which the research organization was given the ownership rights according to the assignment agreement. The company, which had received what it had asked for and paid for, had no problem with this.

- The research organization often get ownership of the project results when the project aims to provide data, information, knowledge, etc. that have minor direct commercial value for the client, or that the client finds no business case for owning. The client is then usually content to obtain free rights to use the results for its own purposes.

  Example: A research group in pedagogy is commissioned by a private kindergarten chain to make a comparative study of pedagogical principles in preschool education. The chain wants to base its preschool offer on research-based pedagogy. It is an important point for them that this is made known to parents, politicians, authorities, and others. They hope this will strengthen their reputation and give them a competitive advantage, even though the competitors can acquire the same information. On this basis, the kindergarten chain finds itself best served by the research organization keeping the ownership of the results of the project and settling for a free right of use.

- The research organization usually retains the ownership rights to patents, technology, software, etc. (background IPR) and equipment that the organization has *prior* to the assignment, but which is made available to the project, and which may be further developed or adapted to the client's needs (usually and if needed, the client gets licences to use this, subject to detailed agreements).
- The research organization usually get the ownership of scientific equipment and 'tools' that have been acquired or developed and used by the researchers in the assignment.
- The research organization usually get a free right to use general knowledge and know-how acquired during the assignment.

In many commissioned projects, the client pays only part of the costs (the research organization can contribute with their own efforts, public or private funding bodies can sponsor the project, etc.). The research organization can also contribute ideas and concepts, IPR, use of special equipment and laboratories, etc. which for one reason or another are not charged to the client directly and which are also not naturally included in the overhead costs. All this may form the basis for the client and the research organization to share or in some sensible way divide the rights to the results of the project, or that the client pays extra for the use of the results in the form of royalties or the like.

When the client is a *non-profit organization*, the starting point will often be different. When there is no commercial justification for the assignment, it is generally less important for the client to have ownership of the results, their interests may be adequately taken care of through further agreed user's rights. Many non-profit organizations, especially public ones, are also generally concerned that research results should be open to all, including the results of their own projects. The responsibility for this is often best taken care of by the research organization, i.e. a university, college, or research institute. The results should then be owned by the research organization while the client acquires the necessary user's rights.

The issues highlighted above are of an inherent business nature. The reason for going into them here is to point out that ethics is also linked to all trade and commerce, and that the parties have much to gain if these interactions from both sides are governed by ideals such as honesty, openness, fairness, etc. The way in which negotiations are conducted, for example, can be particularly decisive for the further climate of cooperation between the parties. Rather than hammering in their own 'demands', the parties should probably rather strive to understand each other, meet each other and show generosity in matters that are of little consequence to themselves, but great consequence to the other party. Solutions that both parties perceive as reasonable and fair can also inspire researchers and have a decisive impact on their motivation and ability to innovate. The examples above also show the diversity and to some extent the complexity of the considerations that should be taken into account, and that in contract negotiations it is generally a question of weighing business matters against non-profit and societal considerations if one wishes to find solutions that are optimal for both the researchers and the client and for the use of national research resources.

Some researchers believe that ownership is essential in order to prevent results from being misused in some way, such as when for various unacceptable reasons a client distorts or seeks to hide important parts of the results, or when newspapers refer to research results in a way that is more suited to selling newspapers than to providing verifiable information. Both the concern and the way of thinking are understandable, but to some extent, it is also possible to prevent certain types of misuse by including some special provisions in the assignment agreement, possibly as a condition for relinquishing ownership rights. For example, one can agree:

- That the researchers themselves, subject to further agreement, may publish information about the project and the results (this is a common provision in commissioned research; see 9.4.4).
- That in certain contexts the client cannot state the names of the researchers or the research organizations without these having approved the use in advance. Examples may be that names cannot be used without prior approval in advertising and marketing, in applications to public authorities and financial institutions where the research results are used in the client's argumentation, etc. Protecting the name of an institution in this way is a common solution.
- That the entire assignment report (without changes) must be attached or submitted if the client intends to use it in documentation to the public authorities, etc.

Such a provision will also strengthen the credibility of the project results when they are used as a basis for decision-making.

- That revised versions or summaries of the assignment report in certain cases must be approved by the researchers. This is a sensible measure to ensure the quality of new editions of the report and can prevent the client from being harmed by reproducing or disseminating the research reports in an inaccurate way.

Support for the last two bullet points may also be found in national copyright laws; see Chap. 17.

Which of these or other provisions it is relevant to include in the agreement depends on the circumstances. In assignments for clients one knows well and has a trusting relationship with, there is less need for these kinds of precautions. The same if there is a low risk that something may go wrong for other reasons. The internal procedures of the research organization may have provisions on this.

Most clients will probably not have problems accepting and understanding this type of minor restriction on ownership rights, not least when they are well-founded and there is an understanding of how they can contribute to creating a trusting relationship between client and researchers and give the results of the project increased credibility.

9.4.3   Provisions on Reporting to the Client

It is common to agree that the researchers must account for the work and the results in one or more reports addressed to the client. Depending on the nature of the assignment, this can be varied, for example by reporting on the progress in regular notes which are then summarized in a final report. Sometimes it is also agreed that the project's administrative matters (organization, staffing, finances, time schedule, lists of reports, meetings, etc.) must be reported separately. An assignment report is not the same as an article in a scientific journal. It must be adapted to the nature of the assignment and the client's needs. If the report is to be used by non-specialists, it should, for example, be formulated in such a way that readers can understand it.

Regardless of the nature and scope of the assignment, it is important to keep in mind that the assignment reports (both the administrative and the scientific one) are the researchers' most important means for documenting that the work has been carried out in accordance with responsible research practice. For research ethics reasons, the following information must always be provided:

- Information about the relationship with the client, such as:

  - Who the client is, and who has financed the project (the project may have been supported by others than the client).
  - Information on the researchers' and research organization's previous and present relations to the client, especially when this may be relevant for others' assessment of the work (more about this in Sect. 16.8).

– Information about who has contributed to the work and what the contributions consist of (in particular, the client's possible participation in the work, contribution with information and data, equipment and materials, etc. must be stated).

- Information about any instructions or limitations in the work given by the client that – according to the researchers opinion – could possibly give a skewed, incomplete, inaccurate, or uncertain picture of what one has studied. The information should be factual, accurate, and objective. This should be mentioned both in the report summary and in more detail in the places in the report where this information is relevant, such as in the description of the scope of the assignment, the choice of methods, and in the discussion of the results.

This information also belongs where it is relevant in project descriptions, working notes, sub-reports, etc. It must always be included in the final report and in publications and open presentations of the project. The assignment agreement must not contain provisions that prevent this.

Sometimes the assignment is part of a larger project, in which other research organizations and perhaps the client's own experts also participate. It may then be agreed that each participant submits reports on their own contribution, which one of the participants or the client compiles in a final project report. In these cases, the assignment agreement should have a provision to ensure that each participant's contribution is used in a proper and truthful way in the final report. This can be arranged for in different ways, the simplest of which is often to let each participant write free-standing reports of their own work, which the writer of the final summary report refers to according to common practice in science. Another solution is that the participants, based on the sub-reports, write a joint final report where the contributing researchers are co-authors according to common research practice (see Sect. 16.6).

9.4.4 Provisions Concerning Publication

Provisions Concerning Confidentiality Versus Public Access
Information about research projects and their results should, as a general rule, be publicly available to everyone. The reason, which has its basis in both research ethics and research policy, is discussed in more detail in Sect. 16.2. In commissioned research, however, some of the arguments for publication do not apply or are less important, and for some assignments, publication may be unreasonable or even harmful (see later).

In principle, most people will probably find it acceptable that clients who pay for R&D work want to keep the results to themselves, and that a research organization accepts assignments under such conditions. On the other hand, codes of conduct for research require researchers to be *open* about their research. Openness in commissioned research is essentially realized as the researchers describe their work, results, and all factors related to the assessment of their neutrality, impartiality, and independence in a truthful and comprehensive manner in their reports to the client. *Publishing* is a tool to make this *available to everyone*. While it is difficult to imagine situations where anything could justify the requirement for openness be deviated from, the question of publication may depend on the circumstances.

The research community includes researchers who work under very different operational conditions, with researchers in the business sector and at universities as two extremes. National and international guidelines for research ethics should in principle take this into account. Guidelines that can apply to all researchers make collaboration easier (research collaboration between academia, the institute sector, and the business community, for example, is a priority in many countries). Equal practice also makes it easier for non-scholars to understand what they can expect from research. In practice, it is almost only when it comes to the question of making research results publicly available that what is right for one type of research is not always right for another. Different national and international guidelines for research ethics take this into account to some extent, but are not always explicit and do not provide uniform advice on the publication of commissioned research. Looking broadly at different guidelines and practices the main principle is clearly that the results of research should be published and accessible to everyone, but that this principle can be deviated from when there are good reasons for it. This is sufficiently nuanced to apply to all types of research. The guidelines are otherwise usually clear that researchers and clients must *agree in advance in their assignment agreement on all issues related to the public's access to information about the assignment*. Detailing here can prevent misunderstandings and disagreements.

However some researchers are not completely free to agree on whether the assignment may be published or not:

- Some commissioned research projects are part-financed by funding organizations that set publication as a condition for supporting the projects, unless special circumstances make this unreasonable. The legitimate reasons for not publishing a research result are then often related to results of commercial value (more on this in Chap. 18). To get access to financial support, one must adapt to these conditions.
- In many countries, national law gives researchers at some institutions (primarily universities and colleges) the right and duty to publish the results of their own R&D. In such cases they may not be totally free to agree with a client concerning publication, but have to follow provisions in the national legislation and institutional regulations.

In negotiations concerning confidentiality versus public access to the results of a commissioned research project, many clients will argue that the project should be confidential. However, on the basis of what has been discussed above, the researchers should encourage the client to the contrary: as a starting point, assume that both the project and the results can be published, and then assess whether there are information and results that for various reasons should or must be kept confidential. The main principle in the assessment should then be to weigh the advantage of publishing against the disadvantage for the client and others. If the value of public access is minor, publishing becomes less important.

With such a mindset, one should for example especially strive to *publish*:

- Results with scientific news value and significance.
- Results that seem important as a basis for political and administrative decisions in society.

What one, on the other hand, cannot or should *not publish* is, for example:

- Information that for business reasons the client does not want to be made public, or where the timing of publication is important.
- Information that is primarily only of interest to the client or that has little scientific significance. Nor can it be right from an ethical nor a practical point of view to burden the news arenas of society or research with matters of insignificance. However, some will think differently about this, and the online community has also made it less important than before to set minimum requirements for quality and importance when something is to be published.
- Information and data that may be harmful or burdensome to individuals, groups, or society in a broader sense (this refers to information beyond what is required by law to remain confidential).

In the negotiations between the researchers and the client about publishing, the researchers should also point out that *publications are important for their careers*. Experienced clients are well aware publications means a lot to researchers, and therefore often go to great lengths to find solutions, even if their own business has no interest in publishing. In technical-scientific assignments for industry, for example, one often arrives at solutions where general knowledge gained in the project is sought to be published, while the results that are specifically related to the company's products and processes, which should not be revealed to competitors, are exempted from public view.

The matters discussed above apply to R&D assignments. If the assignment does not include R&D, many of the arguments for publication discussed above are of little relevance.

Provisions Concerning the Timing of Publication

For many clients, it is important that the results of the project are not published until after a certain time or at a specific time. This may, for example, apply to:

- Results that may lead to an invention of commercial value, and which the client wishes to protect by patent. One of the prerequisites for obtaining a patent is that the invention is not published before applying for a patent. Once the application has been filed, the results may be published. The patent application itself is also published by the patent body after a certain period of time. Patenting is discussed in more detail in Sect. 18.3.
- Results that have commercial value but which the client for various reasons wishes to protect through secrecy rather than patenting. The reasons may be that the results are not patentable (but still to the client's great benefit and of interest

to competitors); that patenting does not provide adequate protection; that patenting would be too expensive, and more. Sometimes the client only needs a headstart over its competitors. In such cases the results of the project, or parts of them, can in principle be published after a while. However, most clients will probably prefer long-term secrecy of business-critical technology and solutions.

- Results that the client wants to evaluate before they are published. This can be particularly relevant if the results may lead to public attention or debate, which the client want to be prepared for.
- Results that are so important that the client wants them assessed, quality assured or verified by their own experts or by external professionals before they are published.

Provisions Concerning the Method of Publication

A research result can be published in very many ways. It is therefore important that the assignment agreement also contains provisions on *how* the results may be published. Most R&D assignments are reported to the client in reports adapted to the client's needs. The simplest form of publication is to publish the reports openly on the internet, on the client's and/or researchers' websites. But sometimes it is more complicated than that. For example, the client may need the results to be presented on a certain occasion (at a professional conference, press conference, etc.). They may also want to present the results in a different format than the researchers' reports, for example, adapted to a specific group of users or stakeholders. In order to avoid misunderstandings and disagreements in the final phase of an assignment, such details should as far as possible also be agreed upon in the assignment agreement. It should furthermore be agreed whether the researchers may present the results, or parts of them, at conferences and in scientific journals and who the co-authors of such presentations and publications should be. A clear and detailed agreement on this is especially important if several parties share ownership of the results.

9.4.5   Provisions Concerning Guidelines for Research Ethics

When two parties are to enter into a collaboration, it is seldom possible or appropriate to make an agreement that explicitly discusses all matters and all situations that may occur. In order for the client to have a certain degree of security in matters that are not specifically mentioned, it is therefore common to insert a number of provisions of a more general nature. For example, that 'the researchers shall carry out the work in a professional manner in accordance with recognized norms, standards and good practice for this type of work'. Following the establishment of fairly comprehensive written codes of conduct for research at both national and international level, one also sees general formulations of the type: 'The assignment must be carried out in accordance with the guidelines for research ethics … [further specified]'. Any clarifications or deviations from these guidelines, which the parties find appropriate and ethically justifiable based on the circumstances of the assignment, can then also be included in the assignment agreement. As discussed in Sect. 2.4.5, such clarifications are particularly important when the assignment does not concern R&D, but for example consultancy work.

9.5   The Relationship Between Client and Researchers

Most researchers perceive the collaboration with the client as demanding but interesting and educational, something which provides meaning and often a boost of excitement. Occasionally, however, they may also experience situations where the collaboration is put to the test, for example, if the client's and the researchers' interests seem to diverge. In such cases, research ethics can act as an important guide to think and act wisely. Some typical situations are discussed below.

9.5.1   The Balance Between Loyalty to One's Own Organization and Loyalty to the Client

Researchers that work on assignments for a client may be faced with the following loyalty problem: Their primary responsibility is to carry out the project in accordance with the assignment agreement for the benefit of the client. At the same time, they are of course also obliged to look after the interests of their own organization. Conflicts can then arise if the client's interests seriously deviate from the interests of the research organization or deviate from the researchers' own opinions about how things should be done.

> Example: During work on a commissioned project at a research institute, the project manager (PM) finds that instead of performing a special analysis in-house, it will be significantly cheaper for the client if the analysis is outsourced to a subcontractor. The institute, which has spare capacity to perform the analysis, will then lose this job and the income that follows from it. Whom should the PM be loyal to? She chooses to outsource the job. Researchers with a stronger sense of duty towards the assignment than towards their own organization are not so unusual in institutes with many commissioned projects. In this case, however, the institute and the client have a contract where the institute's standard prices for all the planned tasks are listed and agreed upon. The client, therefore, does not expect the PM to look for cheaper subcontractors. There is also a certain connection between the ideas on which the project is based, which the institute has come up with for free, and the total scope of work that follows. In this case, the PM's choice can therefore not be justified.

Whenever there is a doubt whether one should act in the best interests of the assignment/client or one's own research organization, the decisive factors will always be the provisions of the assignment agreement (legal obligations) and what the client can otherwise reasonably expect (ethical obligations).

9.5.2   Financial Honesty Towards the Client

Honesty in commissioned research not only concerns integrity in science but also honesty in the use and reporting of funds made available for the project. The financial provisions for the execution of an assignment are always agreed upon in the assignment agreement, and usually, one of two payment methods is chosen:

- *'Fixed price'*. The researchers undertake to carry out a detailed piece of work or deliver a specific result for a fixed price. If the job is completed before all the money has been spent, the excess goes to the research organization. If the job is not done when the money has been spent, the research organization must finance the remainder in order to finish the job. Under such an agreement, the research organization takes a risk, which may be compensated for by including unforeseen costs and compensation for risk in the price agreed for the assignment.
- *'Time-and-material cost within an upper-cost limit'*. The research organization undertakes to carry out a specific assignment or to work towards a goal up to an agreed cost limit. The client then pays for documented worktime and other costs. When the agreed, upper-cost limit has been reached, the work is stopped, regardless of how far the work has come. The client can then either terminate the project or pay more for the work to continue. Sometimes the assignment is carried out without an upper-cost limit. In that case, the client usually keeps a sharp eye on the work. Since in this case the research organization's financial risk is moderate, it normally does not receive any profit beyond what may be embedded in its agreed hourly rate. In some cases, however, the parties may agree on a performance bonus for reaching a specific goal or a time bonus for finishing the project earlier than planned.

Both payment methods require researchers to be honest and precise when filling their timesheets. However, sometimes things are done wrongly, as illustrated in the example below:

> Example: A research institute takes on an assignment for an industrial company under a 'time-and-materials' contract with an upper limit of €50,000. The project is going well, and the task is satisfactorily solved by the time €45,000 have been spent. However, the project manager knows that the client will not notice anything if she fills a few more hours on the project timesheet (hours which she spends on other projects she has in progress). The client is therefore sent a project report that answers the objective of the assignment well within the deadline – as well as an invoice for €50,000. He is very happy. The project manager has thus provided the research institute with a 'bonus' of €5,000, which she defends to herself by saying that the skill shown in the work deserves to be rewarded. With such reasoning, it is easy to see that the higher the cost limit the project manager has obtained in negotiations with the client, the greater the hidden bonus will be. The cost limit of €50,000 is estimated by the project manager, and the fact that the project can be completed for €45,000 may just as well be due to the project manager's poor budgeting skills as to her skills in performing the research.
>
> However, many research organizations have experienced that the opposite is more common, i.e. that researchers do not finish within the agreed upper-cost limit, and that instead of requesting more money from the client, they finish without charging hours on the project, and often without informing their leaders. The research organization then ends up

subsidizing the client, often without management giving the go-ahead for it. Both courses of action are breaches of good practice.

This and other forms of manipulation with timesheets and project funds are sometimes trivialized in research communities when it happens at a small scale and may serve the research as a whole. Few call this a breach of trust, breach of contract, or anything worse. Almost all R&D projects are somewhat under-budgeted, and one can say that almost all researchers are in need of more money for their research. Developing a habit of being open and honest when reporting hours spent on different projects must form part of a researcher's basic training, especially in institutes where commissioned research constitutes a major part of the activity.

9.5.3   If the Client Seeks to Influence the Researchers: What Is Undue Pressure?

It is both common and highly legitimate that the client and the researchers have somewhat different views on one thing or another during the planning, implementation, and completion of an assignment. Both parties argue for their views, needs and interests. Good practice dictates that the parties respect each other's opinions. Nevertheless, from time to time, researchers express that they feel *pressured* by the client's arguments and demands. This may typically take place during the negotiation of the assignment agreement, or in connection with any changes in the agreement caused by unforeseen events along the way. It may also happen in connection with reporting, publication, and use of the results. Many are then faced with the question of when the other party's argumentation should be seen as pressure, and where the limit goes for what to accept.

People react to external pressure differently, depending on their mental sensitivity and strength. For some, the threshold for experiencing pressure is very low, so low that clear statements from a client can be perceived as pressure. However, 'pressure from a client' are strong words that are difficult to justify unless there are obvious and significant negative consequences of not doing what the client wants. *Unacceptable pressure* would then be a proper term if the negative consequences are expressed explicitly, are out of proportion relative to the disagreement, or are presented in a threatening manner. Examples of undue pressure can be:

Example 1: An applied social research institute negotiates a research assignment for a public agency. The client's case officer, who has a relevant social science education, wants the researchers to use a specific methodology in one of the work tasks. The researchers, for their part, argue that this approach may lead to a skewed picture of the object of their survey. The case officer disagrees and states that the institute will not get the assignment if they are unwilling to use the methodology he favours. He adds that refusal will have consequences for the agency's use of the institute in the future. The risk of the project being lost if the agency's view is not followed puts considerable pressure on the researchers, who need the income from the assignment. However, it is difficult to see this as undue pressure. The fact that an assignment does not materialize is a fully natural consequence of the client and contractor not agreeing for one reason or another. The statement concerning consequences for future cooperation, however, should be seen as undue pressure because such a consequence is not reasonably justified. It is simply a threat made only for the purpose of putting pressure on the institute.

Example 2: Halfway through an assignment for an industrial company, a research group finds a technological solution that is patentable. According to the assignment contract, the company shall have ownership of any patents. Therefore, by agreement with the client, the researchers make a draft patent application which they send to the company. Here they claim to be inventors, as good practice dictates. However, the professionals in the company, who have initiated and led the overarching project and contributed to the scientific discussions along the way, see themselves as inventors. They inform the institute's project manager that they have erased the researchers' names and entered their own names as inventors. The project manager responds that inventors' rights are protected by national law, and that the institute's researchers obviously fulfil the conditions for being inventors, but that they are willing to discuss whether some of the company's professionals have also contributed to a sufficient extent to be named co-inventors. The company answers briefly that it owns the results and can therefore do as it pleases. They also announce that the project and all other projects they have with the institute (there are several) will be stopped until the institute accepts this. This is an obvious and undue threat because it entails a requirement that researchers waive their statutory right to recognition as inventors (there are no significant economic benefits to this for either the researchers or the institute). The threat to stop all projects is of course also made just to put pressure on the institute – far beyond the limits of reason and acceptability.

Researchers who become subject to undue pressure should raise the issue at a high level in their own organization at an early stage. There are two main reasons for this: Conflicts are often solved when the client's and the contractor's top managers talk together – leaders in the business and public sector are concerned with having good relations with the research community. When a client threatens consequences, there are often consequences of a financial nature (the halting or loss of projects or payments). This is the management's concern.

Giving in to pressure from a client when it implies violating laws, rules, and research ethics norms is unacceptable. Giving in for other reasons (for example to avoid a conflict, to avoid losing commission income, etc.) is up to the research organization on a case-by-case basis.

9.5.4   If the Contractor Seeks to Influence the Client: What Is Undue Pressure?

In order to exert pressure on another, one must have a means of applying pressure. In commissioned research, the client's obvious means of pressure is withholding payment, while researchers and research organizations have little access to such measures. Therefore, it is usually researchers who express a sense of pressure from the client. But the opposite can also happen, and one means of pressure that researchers can use on clients are threats *to blacken the client's reputation* through accusations or raising suspicions about the client's intentions, behaviour, way of operating, professional competence, products or services, etc. This is of course unacceptable, and especially severe if the statements are not objectively justified, or are simply untrue, distorted, or exaggerated. An example can illustrate this:

> Example: A research group at a university college has developed a modified life-cycle analysis for calculating energy consumption in the production, use, and reuse/destruction of various products which they believe is more accurate than commonly used life-cycle analyses. A local industrial company, which has long collaborated with the college on master's degree projects relevant to industry, hears about the new method. They are in the process of developing a new product that they believe will be more environmentally friendly than competing products and decide to engage the college researchers to conduct an analysis of the product using their new method. However, the result does not turn out the way the company expects. The analysis concludes that the energy consumption from 'cradle to grave' is significantly greater than what the company itself has arrived at, and slightly higher than competing products. The researchers' explanation is that their life-cycle analysis includes some new energy contributions that have previously been ignored and that they weigh the energy components in a more correct way. After some investigation, the company's own experts find that the new method systematically and without scientific justification favours the use of certain materials and manufacturing processes. They also discover that the researchers have obtained data for energy consumption in raw material production from scientific literature, which is not representative of the actual energy consumption in modern, industrial production. The researchers, for their part, argue that these objections are only being made because the company is dissatisfied with the result. They state that they have finished the work and will publish the results. The company then informs the college that the work, according to their opinion, does not meet academic quality standards and that they cannot approve the assignment until it has been verified by a neutral third party of their own choosing, a university abroad with well-known specialists in life-cycle analyses. The college researchers do not like this. They repeat that this is only being done to keep the results of the analysis out of the public eye. They claim that the public has a right to be aware that the company is trying to gag the researchers and conceal the fact that the new product is not environmentally beneficial. If the assignment is not approved and paid for immediately, they will therefore go public with this in the media. They will also refrain from future collaboration with the company on master's programmes. This threatens the company's reputation and puts it in a difficult situation. It is dependent on external partners to implement its R&D strategy and needs to recruit professionals, preferably locally. Conflicts like this are eagerly blown up in the media, and both the media and most people find it easier to believe in researchers than industrial companies. In this case, the researchers are obviously putting unacceptable pressure on the company because their approach is directly threatening and there is a professional disagreement that the company seeks to resolve in a constructive way by letting other, neutral experts verify the work. At this point, however, the college leaders intervene and make their researchers understand that they are not served by proceeding as they are threatening to do.

From an ethical point of view, it is just as unacceptable for researchers to exert undue pressure on a client as for the opposite to happen.

9.5.5   If the Client Requests Changes to the Assignment Report

Many times it is appropriate, or stated in the assignment agreement, that a draft of the project report shall be sent to the client for comments and quality assurance of facts, etc., and to ensure that the client will formally approve the delivery of the service. It is not uncommon for the client, in addition to pointing out factual errors, also to request changes so that matters that are essential for the client's use of the results are better presented, more clearly explained, more fully justified, etc. Compliance with this is rarely a problem. However, what the researchers *must never accept* is a request or order from the client to unduly withhold information or describe elements of the project, its implementation, and results in an untrue, skewed, incomplete, or inaccurate way. Researchers' credibility in society is based on the certainty that they do not enter into ethical compromises.

Clients who request changes to an assignment report usually have good, scientific reasons for it:

Example 1: Over time, there have been indications that a specific treatment used at a private health service company does not work as intended. The company wants to take a closer look at this and starts by engaging a neutral research institute to conduct an independent review of the treatment and to clarify the facts in the specific cases where the treatment seems to have failed. The researchers make it a condition that the project report is published – the results may be of public interest. The work is carried out and the researchers conclude that the treatment has worked poorly for certain patients. They believe that the problem is of such a nature that it will have consequences for the company's use of the treatment. To be sure that the facts are correct, the researchers therefore send a draft report to the company for comment. The company, which has its own qualified professionals, carefully reviews the draft and ends up with several objections. In addition to some factual errors that the researchers correct, a discussion arises concerning the reliability of the information that the researchers have found about several of the cases where the treatment failed. The parties also have different views on the cause of the problems, i.e. whether they are due to a general system error or only poorly performed individual treatments. Even if the parties disagree, they respect each other's professional views and agree that the researchers must complete the report as they see fit, but also briefly mention the client's divergent *professional* views. This approach had several positive elements. The actual errors were corrected, the company's professionals were shown respect, the principle of contradictions was followed, the parties' contradicting professional views became evident, and the uncertainty related to some of the findings was well elucidated. As the company would in any case have to oppose the researchers' report on the points where there was professional disagreement, this was also a simple and practical solution, both for the company and for the readers of the report.

However, this case could also have had other outcomes:

Example 2: The researchers could have asked the client to check whether the facts were correct, taking into account what they thought were relevant corrections, and then completed and published the report without including or informing about the deviating assessments of the professionals in the company. This is probably a fairly common approach. The company asked for a neutral report and got it. If the company has objections of a professional or other nature, it may then publish its views later in a manner it deems appropriate. To this, it can be said that the researchers thus published a report to which they knew there were qualified, professional objections. One can ask whether it is good practice to keep this hidden. The crucial point here is of course whether the company's professionals are well qualified and are able to argue objectively.

Example 3: The report could also have been completed without checking the facts with the client and without asking for comments (there was nothing in the assignment agreement that required the researchers to send a draft to the client for comments). In this case, the report would then have been published with indisputable factual errors, and with information and conclusions of an uncertain quality. Researchers should do their utmost to ensure

the quality of their work. If there is reason to believe that contact with the client can contribute to this, it should be done. Contact with the client is inherent in all commissioned
activities, and the public's view of the researchers' integrity and objectivity should not be
undermined by natural and well-founded communication between client and researchers.

Then, of course, it also happens that some clients request changes to the assignment
report without having acceptable reasons for it. This often comes about due to ignorance, and most clients then accept the researchers' views when it is explained. If
not, the conflict should be handled at higher levels in the involved organizations and
the research organization should consider taking steps to protect its reputation and
its researchers' professional and ethical integrity.

A distinction must thus be made between clients who have good reasons for
requesting changes in the assignment report and those who do not. As in the examples above, many clients are sufficiently competent to assess the academic content
of a research report, or at least parts of it, and in principle their views should be seen
as an element in the quality assurance of the work (researchers can also make mistakes or perform low-quality work). Another argument for cooperation on assuring
the quality of a scientific assignment report is that on their own, most clients will
have difficulties disputing the report after its publication – let alone being heard and
believed, even when their own experts are well qualified in the field and the objections are justified.

9.5.6 If Client and Researchers Have Different Views on the Interpretation or Use of the Project Results

It happens from time to time that clients and other users of the project results assess
the importance and use of the results in a different way than the researchers. In cases
of great public interest, this can receive considerable media attention and lead to
debate – not least in cases where there are different political or ideological views in
society:

Example: A group of traffic researchers is commissioned by an urban municipality to study
the connection between road tolls to enter the city and the use of private cars. It is agreed
that the assignment report shall be published. In their work, the researchers use a computer
model that calculates the traffic based on different assumptions. They find that by increasing tolls considerably during rush hours and reducing them at other times of the day, one
can influence driving patterns and car choices so that CO_2 emissions from urban traffic are
reduced by 20%. However, the city's political leadership choose not to do so. They are then
attacked by the opposition parties who want high tolls, and who refer to the researchers. The
researchers also enter the fray to argue for what they see as their 'solution'. They are backed
by other researchers with less relevant competence who take this as another example of
arrogant politicians who only use research results that support their own political views.

For the client – the city administration and political leaders – however, the reality is different. They had not asked the traffic researchers to find a solution on how the city could
reduce its CO_2 emissions. The purpose of the work was to get better data related to a small
part of the city's CO_2 emissions problem, which they needed to evaluate different considerations and measures against each other. They wanted, for example, to avoid too large tolls

that could lead to social differences in the use of private cars. They were also interested in considering alternatives, such as making it cheaper to travel by public transport, banning certain car types or cars in certain zones, and more. They also wanted to weigh the measures on the traffic side against other ways to reduce CO_2 emissions.

In this example, the client had the broadest competence to assess the use of the research results. The researchers' user competence was 'narrower'. They overestimated the importance of their own research results and had insufficient knowledge of the circumstances related to their use. This is not unusual. Research is to select a small part of reality and study it with scientific methods under well-defined assumptions and approaches. This does not provide good training in assessing the issues encountered in the public and business sector, where it is always a question of weighing many different factors and considerations against each other based on deficient and partly uncertain facts. In an ethical context, it is just as important to have respect for clients' and users' broad societal competence as it is to have respect for other researchers' specialist competence. However, experienced researchers can of course over time develop user competence in specific areas through cooperation with users, not least by working for the same client over a longer period of time.

To balance the example above, one can also envisage situations where for *unacceptable* reasons a client ignores research results. However, before concluding on this, the issue should be discussed with the client to seek insight into the client's situation and reasoning. Some clients may refuse to be open about this, but many will probably see this form of feedback as a good investment in future collaboration.

9.6 The Risk of Losing Neutrality when Working on Commission for a Client

As discussed in Chap. 7, no researchers, regardless of whether they work on their own projects, in collaborative projects or on commission for others, are completely neutral in all contexts. Neutrality is an ideal that no one can fully realize in practice. In commissioned research, it is the relationship with the client that particularly threatens neutrality, i.e. the fact that the client decides or participates in deciding the topic of research and pays for the work. In long-term collaboration with the same client, there is also a risk that the researchers begin assuming the client's interests or becoming dependent on the income the assignments provide. None of this can be called immoral, but it can make it difficult for the researchers to be objective and balanced in their research, and others may view the research as less reliable because of the longstanding connection with the client.

These problems must be met with *openness*. The credibility of a commissioned research project can be strengthened by being open about the relationship with the client, the purpose and scope of the assignment (including any relevant limitations in the scope), how the project is carried out, the methods used and the reasons for using them, etc. Transparency will also motivate the researchers to let their work be characterized by objectivity. In cases where it is important that the public at large has confidence in the results of a commissioned research project, the openness must be directed toward the public, i.e. the research report must be published or in other ways made accessible, as discussed in more detail in Sect. 9.4.4.

Research institutes with many commissioned research projects face a particular challenge when it comes to being perceived as independent in relation to clients. Transparency about each project is then rarely enough, especially not when it comes to research on politically controversial areas or areas of interest for long-term clients and clients who represent large revenues for the institute. These institutes should therefore work fundamentally and systematically not to be inappropriately influenced by the clients and to be so transparent and honest about their activities that the public gains confidence in them. The first is a question of training all employees to be useful partners for the clients while maintaining their integrity and keeping a 'distance' from the client that makes it possible to deliver trustworthy results of high scientific and research ethical standards. The latter has to do with establishing an organizational culture that both practices openness about the institute's mode of operation and focuses on the implementation of responsible research ethics in the organization as a matter of course. Independence and objectivity must be demonstrated through action. This can be seen as an element in a definition of 'responsible *commissioned* research practice'.

From time to time, researchers working on commission for a client will be accused of serving the clients' interests in an inappropriate manner. When the allegations are serious, the best defence may be to treat them as suspicions of violations of research ethics norms and investigate them formally, as discussed in Chap. 23. If the accusations are unjustified, this is a credible way to be 'acquitted'. Highly trustworthy, *external* and neutral experts should then be engaged to investigate the case. The fact that a research organization does this as a matter of routine is in itself documentation that the ideals of objectivity and independence are set to a high standard.

Chapter 10
Other Forms of Research Collaboration Between the Research Community and Society

10.1 New and Old Measures to Involve Society in Research

Most people believe that research serves society well, but there seems to be an increasing focus on how researchers' and research organizations' contributions to society can be optimized. In this context, special emphasis is placed on the utilization of technological innovations in data processing and communication. Another approach that has been prioritized in recent years – which seems to be both scientifically, ideologically, and politically motivated – is to open research to the public in new ways and bring researchers, users, and the general public closer together in what is called 'Open Science' (Open Science, n.d.). The idea is that broader and more open collaborations will contribute effectively to the sustainable development of society and that the public's confidence in research will be strengthened. Open science is a wide-ranging concept that is currently under development and which research authorities everywhere are now stimulating in various ways. Some of the measures are discussed elsewhere in the book (see responsible research and innovation, which was discussed initially in Chap. 6; open access to research data, etc. which is discussed in Sect. 12.5; open access to publicly funded research publications, which is discussed in Sect. 16.9.2). This chapter deals with two forms of open science: Project collaboration with potential users of the research results (an established form of collaboration), and collaboration with the general public (a form of collaboration in development).

© The Author(s), under exclusive license to Springer Nature
Switzerland AG 2023
D. Slotfeldt-Ellingsen, *Professional Ethics for Research and Development
Activities*, https://doi.org/10.1007/978-3-031-25484-0_10

10.2   Projects with User Participation

Both nationally and internationally, authorities have long had instruments to stimu-late research communities to involve potential users of the research results in the projects. The term 'user' is broadly defined: business enterprises, public bodies and authorities, interest groups, associations and voluntary organizations, individuals, etc. The scope and content of the participation can vary greatly and include all or part of the project work, from the idea and planning through implementation to the use of the results. Basically, one can look at these as collaborative projects in which the parties' motives for participating and their contribution to funding, planning, and implementation, can vary more than would be usual in collaborative projects between researchers only. A research organization will always be responsible for the project, while the participants' rights and obligations will otherwise be defined in an agreement relevant to the collaboration. Much of what has been said in previ-ous chapters on cooperation agreements and ethical issues in connection with them will be relevant here as well. However, the diversity of projects with user participa-tion means that the ethical issues associated with the collaboration vary consider-ably. Three examples:

• Some users may have extensive professional competence (perhaps also relevant research competence) and experience in collaborating with researchers. The col-laboration can then take the character of a collaboration with other researchers. The ethical issues to be dealt with here will typically be related to the partici-pants' scientific and ethical responsibilities in the collaboration, co-authorship when the results are published, co-ownership of the research results, etc.
• Some users may lack competence and experience in collaborating with research-ers. In such cases it will be important to establish an understanding of and agree-ment on the premises for the collaboration and the expectations of the result. The users may also have a limited understanding of the code of conduct for research, so it will then be necessary to clarify the professional and ethical requirements for research and find practical forms of collaboration that allow the users' contri-butions to satisfy these requirements.
• Some users may be very resourceful and influential, others not. In addition to what has been mentioned above, it will then be important to ensure that the resourceful users' interests do not affect or utilize the research and the results at the expense of the less resourceful.

In all projects with user participation, in the planning phase it will also be especially important to clarify possible conflicts of interest, both between the individual users and the researchers and between the users.

The variation in user participation and project content makes it unsuitable to set up general ethical guidelines for these types of collaborations, but for certain subject-specific projects one may be able to find ethical and practical guidelines for user participation.

10.3 'Citizen Science'

In recent years, researchers in many countries have become increasingly interested in exploiting the potential that lies in engaging laypersons in the definition, planning, and/or implementation of research projects. Here the term 'layperson' includes individuals, groups, or organizations that do not have relevant research expertise. For example, engaging people interested in nature conservation in collecting information about pollution or observing and reporting damage to nature, which the researchers then use as input data in their projects. Or engaging amateurs with metal detectors in archaeological fieldwork. Laypeople who participate in this way are often enthusiastic and interested individuals or groups who can contribute great work efforts. However, it is important to note that citizen science is research conducted by competent researchers, with the participation of laypeople, and not research carried out by laypeople without the participation of qualified researchers. A more comprehensive description is provided in a note by European Citizen Science Association (ECSA); see ECSA (2020).

Citizen science spans a diversity of laypeople's involvement and many forms of collaboration between researchers and the society. It has greater potential in some fields of research than others. When laypersons participate in research, a number of new and unusual methodological and ethical issues arise that the research community must address as new forms of citizen science develop. However, a general challenge is to ensure that participating laypersons know and follow relevant guidelines for research ethics. This requires special training, guidance and monitoring of the participants.

In many citizen-science projects, laypersons will be engaged in acquiring local knowledge, or culturally rooted and experience-based knowledge (sometimes referred to as 'traditional knowledge'). These knowledge bearers are usually unfamiliar with research and collaboration with researchers. The dialogue with them can therefore be demanding and require good communication between researchers and participants from the planning of the project to the implementation of the results.

Laypersons participating in research deserve respect. The researchers should:

- First of all, listen.
- Let the laypersons experience real participation (and not the experience of being 'used'); give them something in return for participation (other than money and gifts); credit them duly for their participation.
- Make sure that the use of the research results is perceived positively in the local communities from which the traditional knowledge is obtained.

The methodological and ethical issues (often interrelated within citizen science) that are specific to the laypersons' participation in research, are largely specific to each project and cannot be addressed in general.

References

European Citizen Science Association. (2020). *ECSA's characteristics of citizen science: Explanation notes*. Retrieved September 21, 2021, from https://ecsa.citizen-science.net/wp-content/uploads/2020/05/ecsa_characteristics_of_citizen_science_explanation_notes_-_v1_final.pdf

Open Science. (n.d.). *Wikipedia*. Retrieved September 21, 2021, from https://en.wikipedia.org/wiki/Open_science

Chapter 11
Research and Other Activities Based on Sponsorships and Donations

A major part of the research carried out in the world is funded or part-funded through sponsorships and donations from public and private sources.

In general, the large *national and international research-funding organizations* (national research councils and research funds, the EU Commission, private foundations, etc.) invite research communities to apply for funding for projects, equipment, research fellowships, etc. within defined research areas. The applicants must then compete for financial support (sponsorships, grants, scholarships). Although these organizations always impose certain conditions for this support (thematic, collaborative, administrative, financial, etc.), it is always the researchers' own project proposals that are considered and possibly supported. The results, therefore, belong to the researchers and the research organizations, although there may be conditions attached to their use and availability. Because the criteria for receiving support are transparent and clearly defined, and the work is expected/required to be carried out in accordance with relevant codes of conduct for research, there are generally few ethical concerns when applying for support from such funding sources.

It is also common for researchers to seek support from potential *public and private users* of the research. Sometimes public research-funding organizations also set such co-funding as a condition for supporting a project. The sponsors then usually receive certain agreed benefits from the research, such as usage rights to the research results.

Some researchers and research organizations can, often without asking, also receive donations for research from wealthy *individuals, organizations, and companies* who want to give something back to society and who believe in the benefit and usefulness of research. While charitable reasons are likely to be behind all such donations, other reasons may also matter. This applies, in particular, to support from the business community and interest groups, where the donor often gives money to develop a specific research area, build a laboratory, purchase certain equipment, support young researchers, support cooperation with other countries, etc. – which the donor may later benefit from in one way or another. A desire for attention,

D. Slotfeldt-Ellingsen, *Professional Ethics for Research and Development
Activities*, https://doi.org/10.1007/978-3-031-25484-0_11

goodwill, and PR can also play a role. The potentially problematic aspect of receiving conditional donations is that the researchers become bound to activities and initiatives that are determined by the donor (the donor's priorities are not always the same as the recipient's). However, it is both common and necessary for researchers and research organizations to adapt to a reasonable degree to the funding opportunities that exist at any given time. A careful reflection on how far to go is nevertheless advisable.

The identity of the sponsor or donor often plays a big role. Many react if a research organization accepts support from individuals, organizations, or companies involved in controversial activities or who have been caught acting in an unacceptable way in one context or another. Likewise, sponsors and donors want to stay away from researchers and research organizations connected to wrongdoing (this mindset is especially evident in sport where sponsors often flee from athletes caught using performance-enhancing drugs).

All sponsorship funds from users and donations establish a relationship between donor and recipient that can be partially compared to the relationship between a client and a contractor. A degree of the same caution that must be exercised in commissioned research should therefore be exercised when receiving research donations and sponsorships from users.

Chapter 12
Ethical Issues During the Research Project

12.1 From Planning to Implementation

The starting point for the implementation of a research project is the project plan that has been made and the agreements that have been entered into with any partners, funding sources and/or clients, where all research ethical issues of relevance to the project have ideally been taken into account. Planning is one thing, but implementation is something else, and it is through actions that morality is first and foremost put to the test.

A number of special ethical issues related to the implementation of research projects are dealt with in separate chapters, while a few more general matters are dealt with in this chapter.

12.2 Responsible Course of Action when an R&D Project Does Not Go According to Plan

Because R&D projects rarely go exactly as planned, all project agreements with funding sources, clients and partners must have provisions on how deviations from the project plan are to be handled. Any approvals from government agencies will also contain provisions on how changes that are relevant to the approval are to be handled. The most common provision is that the *project manager must report* significant deviations from the project plan, for example, if changes in the project's key personnel become necessary, or if parts of the plan cannot be implemented, cost more or take longer. It is also common that the report includes a description of the discrepancies that have occurred or are about to occur, and a proposal for an adjustment plan that has to be formally accepted by the project stakeholders before the work can continue.

© The Author(s), under exclusive license to Springer Nature
Switzerland AG 2023
D. Slotfeldt-Ellingsen, *Professional Ethics for Research and Development Activities*, https://doi.org/10.1007/978-3-031-25484-0_12

Because all R&D projects are expected to be carried out in accordance with responsible research practice, serious violations of research ethics norms must be treated as deviations that must be reported. The report must not only go to the research organization's management as discussed in Sect. 5.14.1 but also to funding sources and any clients and partners in accordance with the provisions for reporting deviations stated in project and cooperation agreements.

Since deviations from the project plan are common in research, they are sometimes downplayed and the researchers and the project manager choose to continue the project without reporting that something has gone wrong – perhaps in the hope that the problems will eventually be solved. Some may also think that it will be easier to get 'forgiveness' later than to get approval in advance for the adjustments they want to make. Irresponsibility and lack of respect for others can also play a role in those who neglect their duty to report deviations in the project work.

That said, in practice, there must be a limit to which deviations must be reported and which changes must be approved in advance. Minor breaches of good research practice should, for example, be manageable by the project manager alone. The question of where the boundary goes must be subject to discretion on the basis of the provisions of the contract and the circumstances surrounding the project and the situation. When exercising discretion in this regard, it may be useful to define this not only as a matter of contract law, but also as an ethical issue, where it is important to act truthfully, openly, honestly and respectfully towards sources of funding, customers, partners, government agencies and others involved.

12.3   Responsible Use of Project Funds

When works of research cannot for a variety of reasons be executed according to the project plan, there is almost always a need for more money and time to see the project to its conclusion. In such cases it is natural to ask for more money from the funding source(s) or alternatively propose to adapt the rest of the work to the remaining funds. However, the agreement with the funding source(s) may not open for these solutions. Parts of the work must then often be carried out without payment. At universities and colleges, where researchers largely dispose of their own time, carrying on without sufficient funding may sometimes be possible. In research institutes where the researchers' salaries are to a large extent funded by commissioned projects, this is a very limited possibility. Alternatively, the project can be terminated before it is completed, with the consequences this may have in terms of the project agreement.

The problems such deviations lead to can tempt researchers to cheat with the project funds in different ways, especially if they have several projects with external funding going on at the same time. Two examples:

The first concerns manipulating project funding:

> Example 1: A researcher has several projects going on within the same field of research but with different work tasks and objectives. When the work on one of the projects turns out to require more work than planned, he begins writing the extra work on the timesheet of one of the other projects – a project that is running over several years and aiming to build competence in the field, and where the work can be more easily adapted to the means. At the end of the year, he says nothing about this in his annual reports and accounts to his sponsors. He justifies this to himself through the unpredictability of research and that his solution represents a sensible use of his total research funding. This may possibly be acceptable if the work he lists on the timesheets of the multi-year project professionally and contractually lies within the scope of the project, and the annual report for the other project states that the last part of the work has been financed by the multi-year project. However, the right approach would be to discuss the deviation from the plan with the sources of funding or to ask his own leaders for internal funding to finish his obligations. In any case, the reporting to the sources of funding must reflect the truth about how the funds have been used.

The second example concerns manipulating project results:

> Example 2: A research team has a significant research assignment for a group of industrial companies that is running in parallel with a small internal project in approximately the same area, sponsored by a national research council. The project plans are different, but some of the work in the big project intersects with the small one, and vice versa. Both projects have competence-building as a general goal, but specific objectives have also been agreed for the two projects. When the time comes to report to the group of clients and the research council, the team leader chooses to see the results of the two projects as a whole. In the small project, she thus inflates the results by including results from the large project, and she supplements the large one with something from the small project. The real reason for acting in this way is that for various reasons the results are meagre in relation to expectations – something the team leader wants to hide and compensate for by double-reporting some of the results. Although all the funders here are actually quite satisfied (they do not know any better), this is obviously irresponsible research practice. Good practice dictates that one should always state who contributes to the financing of a research project. This did not happen here. Those who, for example, read the reports that were sent to the research council were not informed that some of the results came from an industry-funded project. The reporting shall also make it possible for the funders to assess whether the work carried out is in proportion to the funds used. In this case, the industrial companies were misled into believing that everything was in order, and the research council into believing that the researchers had produced significantly more than expected. The researchers thus obtained unjustified goodwill in the council. Finally, the course of action led to parts of the results being mentioned in project reports twice, in different 'packaging' and wording, without saying anything about it. Another breach of the code of conduct for research.

The risk of getting into financial trouble can sometimes be reduced somewhat by being careful when entering into an agreement with the funders. One can, for example:

- Take into account unforeseen events when budgeting the project, as far as possible.
- Agree to carry out specific and well-defined *work tasks* rather than agree to deliver a more defined result, as far as this is possible.

12.4   Keeping a Daily Research Log

Openness is central to research ethics. In practice, this means that researchers must be open about what they are doing and document what they have done. As an element in living up to this, everyone should keep a daily research log – a 'diary' where work and results are stated. This can be done in many ways. In some research areas and research groups, there is a tradition of keeping very detailed logs. For others, a briefer description may be sufficient. Some research organizations have internal procedures for this that must be followed.

There are several reasons to keep a daily research log:

- The habit of a daily research log is useful for oneself – from time to time everyone needs to go back to what they have done, to check or refresh the memory about something.
- The history of science has many examples of controversies about who was the first to make a discovery. A good, daily research log together with properly stored original data can help clarify such disputes.
- In R&D projects that lead to inventions and patents, collaborators or clients may from time to time end up in disputes about who came up with the invention or who contributed to it. Various research groups around the world may also disagree about who arrived at the invention first. Reliable written research logs and properly stored original data can help secure the interests of legitimate inventors and can sometimes even prevent researchers from encountering trouble by breaking other people's patents.
- On rare occasions, a researcher may be accused of manipulating data in different ways. Reliable written research logs and properly stored original data can then be in support of an honest researcher who has to fight for his innocence in the face of false or unjustified accusations.

For the sake of the last three bullet points, the log must be written so that it can be used as evidence in disputes (numbered pages, entries that cannot be changed afterwards, etc.)

In addition to a daily research log, data and material generated during a research project will be important documentation and evidence of what has been done and found in the project; see below.

12.5   Management and Storage of Research Data and Material

According to the common code of practice for research, the results of research must be *traceable*, *verifiable* and *accessible*. Traceable means that the researcher must document where the data and information come from, what procedures and methods have been used, etc. Verifiable means that the research work and results must be described in sufficient detail so that others can repeat them as far as possible. In certain situations and under certain conditions, original data and material necessary to verify the work should also be open for authorized insight. Accessible means that as far as possible the work should be open to others (see, for instance, ALLEA (2017, Sect. 2.5): '... as open as possible, as closed as necessary ...'). Major research-funding bodies have in recent years especially argued that the vast amounts of research data generated worldwide must be made accessible to other researchers as a source of new research to optimize the return on society's expenditure on research. This has become an important tool in the concept of open science (see Chap. 10). The escalating amount of electronic research data might, for example, be utilized by data-mining techniques. How useful old data is for new research, however, will vary greatly.

Research data is defined in several ways, but the general meaning of the term is texts, figures, diagrams, tables, images, videos, audio files, etc. that are generated in the research projects through collections, observations, surveys, experiments, recordings, calculations, etc. Today, most research data is digitized. In addition comes physical research material such as test samples, finds, objects, original documents, etc., which some also count as research data.

The new way of thinking about research data requires researchers to take action to manage their research data, i.e. to make a *data management plan (DMP)* whenever a new research project is planned. The plan should be based on the so-called 'FAIR Data' principles and should state how the research data generated or acquired during the project will be handled (by collection/creation, processing, analysis, and presentation), and how it will then be stored and possibly made available to others. The FAIR Data principles are: Findable, Accessible, Interoperable and Reusable. These principles were developed in 2014 by a network of professionals, librarians, archivists, publishers and research funders called FORCE11 (The Future Research Communication and e-Scholarship) and later published; see Wilkinson et al. (2016). To finance projects, many research-funding bodies now require a DMP as an appendix to the project plan. However, the requirements for the content and scope of the plan vary, although there are initiatives for harmonization (see Science Europe 2021). Research organizations have followed up by establishing templates for writing a DMP, making arrangements for secure and responsible storage of research data and procedures for others' access to the data. Research Data Management (RDM) has thus become a new administrative activity in research organizations. The researchers' responsibilities and tasks in this area are therefore usually well described.

Many researchers probably regard the DMP as more bureaucratic than directly useful to them. On the other hand, the plan is an easy way for researchers to document that their research is transparent, traceable, verifiable and accessible to others.

References

All European Academies. (2017). *The European code of conduct for research integrity* (Revised ed.). ALLEA. Retrieved September 21, 2021, from https://allea.org/code-of-conduct/

Science Europe. (2021). *Practical guide to the international alignment of research data management* (Extended ed.). Science Europe. https://doi.org/10.5281/zenodo.4915861

Wilkinson, M. D., Dumontier, M., Aalbersberg, I. J., Appleton, G., Axton, M., Baak, A., et al. (2016). The FAIR guiding principles for scientific data management and stewardship. *Scientific Data, 3*(1), Article 160018. https://doi.org/10.1038/sdata.2016.18

Chapter 13
Research Involving Humans

13.1 Research Involving Humans – An Extensive and Particularly Sensitive Field of Research

Research involving humans takes place in many fields, be they medicine and health sciences, biology, humanities, social sciences, law, certain mathematical-scientific and technological fields and more. Today, research involving humans is subject to a comprehensive set of research ethics guidelines that cover a range of issues and subject areas. At the same time, this part of research ethics is so sensitive and diverse that one's own ethical discretion is often put to the test. In some research areas, the work will also be regulated by legislation. This applies in particular to research aiming at obtaining new knowledge about health and diseases and research where personal data is processed. New technologies, such as artificial intelligence and the use of big data based on information about people, their actions, opinions, decisions, etc., also raise completely new and to some extent complicated research ethical questions. Going deep into this is beyond the scope of this general book on research ethics. Therefore, this chapter only outlines the most important ethical principles and statutory regulations for involving people in research.

Everyone who becomes involved in research is affected by it, and sometimes in a harmful or burdensome way. The crucial ethical questions then become how great a risk of harm and burden one is able to justify, how that risk should be weighed against the importance and benefits of the research, and who should make decisions about this. Research ethics answers these questions on the basis of respect for human rights and dignity, people's privacy and people's right to decide for themselves what they should participate in. The starting point is also that research that involves people should always be carried out in such a way that it can benefit individuals, groups and/or society in a broader sense.

Laws and ethical guidelines for research involving humans differ somewhat from country to country and are also regularly adjusted. The regulations are particularly

© The Author(s), under exclusive license to Springer Nature Switzerland AG 2023
D. Slotfeldt-Ellingsen, *Professional Ethics for Research and Development Activities*, https://doi.org/10.1007/978-3-031-25484-0_13

comprehensive in medical and health-related research (an overview of international recommendations related to medical research on humans has been provided by the Council for International Organizations of Medical Sciences (CIOMS, 2016), in collaboration with the World Health Organization). Everyone must therefore familiarize themselves with and follow the relevant regulations in force in their country. The discussion in this book must therefore not be seen as a manual for how to proceed, but as a highlighting of key principles of research ethics and general morality in this area.

13.2   Basic Ethical Principles for Research Involving Humans

13.2.1   Respect for Human Rights and Dignity

The current ethical norms for research involving humans are rooted in modern society's views of human beings, as expressed, for example, in Article 1 of the Universal Declaration of Human Rights: 'All human beings are born free and equal in dignity and rights' (United Nations, 1948). In order to respect this view, a number of practical ethical guidelines and legal regulations regarding research involving humans have been established. One example is the Helsinki Declaration with international guidelines for how to care for and protect humans who participate in medical research. Article 8 states, for example: 'While the primary purpose of medical research is to generate new knowledge, this goal can never take precedence over the rights and interests of individual research subjects' (World Medical Association, 2013). This ethical principle is fundamental to research involving humans in all fields of science.

13.2.2   Respect for People's Privacy

The right to privacy is a human right that is enshrined in Article 12 of the Universal Declaration of Human Rights: 'No one shall be subjected to arbitrary interference with his privacy, family, home or correspondence, nor to attacks upon his honour and reputation. Everyone has the right to the protection of the law against such interference or attacks' (United Nations, 1948). As individuals, everyone has the right to a private sphere which they themselves control without interference from the authorities or other people.

One of the greatest dangers for individuals' private sphere to be violated lies in the misuse of the large amounts of personal information found in public and private archives and registers. Information that are stored electronically, on paper, or as films, photos, sound recordings, etc. Here society has therefore seen a need to protect the individual through national data protection laws. This is discussed in more detail in Sect. 13.9.

Another area where the private sphere of individuals is at risk of being violated is in the misuse of information about them, their activities, opinions, interests, etc. that accumulate to an uncontrolled degree on the internet. This is discussed in more detail in Sect. 13.11. Here, both legislation and research ethics face a number of new challenges that have not yet been resolved.

13.3 What Is Covered by the Term 'Research Involving Humans'

The term 'research involving humans' includes research using:

- *Living* people as research subjects.
- Material from living people (tissue samples, blood samples, etc.).
- People as interviewees, informants and the like.
- People as objects of observation, whether they know about it or are unaware that they are being observed.
- Information about living people obtained from open, confidential or secret sources.

Research may also include *dead* people. It is dealt with separately in Sect. 13.15.

13.4 Requirements for Justification and Assessment of Risk

In an ideal world, no one should be harmed or burdened by participating in a research project. In practice, this will neither be possible nor necessary. A person participating in the trial of a new drug may be willing to accept the risk of minor side effects, as is the case for most drugs. Interviewees may be willing to talk about difficult relationships, even if it may lead to bad thoughts and memories they would rather forget. In practice, it is thus a question of where the line is drawn between acceptable and unacceptable consequences of the participation in the research. In addition, there will always be questions about how important or potentially beneficial the research is in relation to the risk and burdens for those involved. The two questions are related. The more important the research in which people are involved is, the greater the risk and burden they are willing to accept. Good research practice indicates that those who plan a project that involves people, thoroughly investigate and assess these issues. A responsible way to implement this can be to:

- State the aim of the project and consider:

  - What concrete benefit can the participants get from this research?
  - What concrete advantage can other individuals or groups get from it?
  - What benefit can society in a broad sense get from it?

- What new knowledge, information, data or the like can the project provide that is scientifically important?

If the usefulness and importance of the project are low, it should not be initiated.

- Carry out a risk analysis to identify the risk of participants being harmed and burdened.

  - A risk analysis will include assessments of what can happen, how likely it is that the incidents will occur, and how serious each of them is. The thoroughness of the analysis should in practice be adapted to the risks related to each project. For example, it goes without saying that one does not need to put as much effort into assessing the risk associated with an interview survey among students on a non-sensitive topic as one does with a medical study that may cause physical or mental side effects for participants.
  - Risk analysis is a special subject area. The analyses can be performed with different methods. It is important to use a method that is suitable for the issues that are relevant to the research project.

- Use the risk analysis to find ways to minimize the risk for those involved in the research, and to minimize the uncertainty about the consequences of the research. This can be done, for example, by:

  - Omitting research elements of minor importance or benefit.
  - Working more carefully and step by step.
  - Taking special measures that can prevent or reduce the extent of harm or burden that may occur.

- Assess the risks in relation to the benefit and importance of the project, i.e. make a decision on whether the project should be initiated or not.

In the first instance, it is the researchers that plan the project that must make their *own assessments* of this. The big question in these assessments is, as discussed in the introduction to the chapter, where the line can be drawn for what can be considered acceptable risk and burden for the research participants. The question is complicated by the fact that the boundaries differ from person to person – in a given situation, some are more easily harmed and burdened than others. Consideration for those who will be exposed to risk and burden shall permeate such assessments. This means that the risk and burden on the research participant must be low and that projects, where any harm or burden can be long-lasting, are not initiated. It can often also be appropriate to make decisions based on a 'worst-case scenario', rather than a probability-based scenario, as a risk analysis usually is.

Medical and health research involving humans stands out in this context in that the basic principles for balancing risk against benefit are regulated by national legislation. An overriding principle in legislation is that such research should be avoided and can only be carried out under special conditions.

The assessments described in the bullet points above are steps in the research planning that must be thoroughly quality assured and all research organizations conducting research involving humans must have internal procedures in place for doing so. The seriousness of involving people in research is an argument that quality assurance should be carried out by others than the researchers themselves. In some cases, projects involving people in the research must therefore be approved in advance by an external body, subject to national laws and regulations. In those cases where completely new issues arise in connection with human participation in research, it may also be relevant to seek advice from relevant, national research ethics committees or the like.

All researchers who participate in a research project involving humans are responsible for ensuring that these assessments and actions are carried out, and are responsible for their consequences.

13.5   Requirement for Informed Consent from People Involved in Research

The most important measure to protect human rights and respect the dignity and privacy of people who are invited to participate in the research is to ask for their consent to participate, often called '*informed consent*' to emphasize that the participant must be fully informed about the project and the possible consequences of participating. The responsibility for obtaining such consent lies with those responsible for the research project. In all countries, a battery of internal organizational procedures, national and international guidelines for research ethics and legal provisions tell how to proceed in practice in various subject areas and issues. The regulations are detailed and with many special provisions.

However, an overview of the main elements of the regulations can be useful for everyone. Partly as an illustration of how important it is to protect the interests and right to self-determination of people involved in research. Partly as an introduction for researchers who for the first time involve people in research. In the following, therefore, the main principles and rules for obtaining consent to participate in research are reviewed, but some details and special provisions are omitted. However, the reader must familiarize themselves with and follow the regulations that apply at the organization and in the country where the research is to be carried out.

The requirements for a valid consent can be summarized as follows:

- The person giving consent must be *informed* in advance of:

  - What the project is about, and what one expects to achieve with the research.
  - Who is responsible for the project and who is doing the research.
  - How the project is financed.
  - How the person giving consent will be involved, and the extent and duration of this.
  - What information, data, etc. the researchers will obtain or request from those involved.

- Who gets access to the information, and how and for how long it will be stored.
- How the participant's anonymity or confidentiality is maintained. This generally also includes a *promise* that the information about the person in question will be treated *confidentially* as far as the law allows. The promise of confidentiality and its limitations are discussed in more detail in Sect. 13.12.
- Which regulations the project must comply with.
- What rights the person giving consent has to gain access to registered personal information and the rights to make changes.
- That participation is voluntary, and that everyone can withdraw from further participation at any time without suffering personal consequences.
- Any other matters that the participant should be informed about.

The information must be provided in a way that is understandable to the persons who are to give their consent. This means that the information must be adapted to the circumstances and situation of the person involved. For example, children must be informed in a completely different way than adults. The same applies to adults who easily lose the thread in too much and professionally formulated information, and who can thereby accept things they have not really understood.

In some research fields, research organizations and public authorities may have templates for requests for participation in a research project, and for the informed consent from the participant.

- Consent must be *specific* and *explicit*.

  - It must be clear who gives consent, when it is given, and what the consent includes (in terms of scope and time), and the consent must be expressed clearly and unambiguously.
  - Consent is usually given to a well-defined involvement within a time-limited project. However, material and information that the researchers obtain, can often prove to be of interest in later research. An example could be a tissue sample that a patient has given a research group consent to use in a specific research project, which later turns out to be interesting to use in a completely different study. Researchers therefore often want so-called '*broad consent*', i.e. a consent that includes use not only within one project but for a more broadly defined research purpose. This may be acceptable under certain conditions. Broad consent must also be specific. Each research organization has internal procedures for this based on national legislation and guidelines for research ethics.

- The person giving consent must have *capacity and competence to consent*.

  - Competence to consent requires:

    - Ability to understand the information provided about the research project.
    - Ability to understand what the consent entails of advantages and disadvantages (risk of harm and burden) for oneself.
    - Ability and willingness to assess one's own participation on the basis of an independent reasoning.

- Ability to understand what an acceptance entails, and especially that one can withdraw from the participation at any time (on the right to withdraw; see Sect. 13.7).

- Consent to research involving adults who do not themselves have capacity or competence to consent must be given by others (parents, next of kin, etc.) according to legal provisions and ethical guidelines in each country.

  - Examples of adults who may lack or have reduced competence to consent are people with dementia, mental handicaps, physical or mental illness that reduces their ability to understand and reason, and people under the influence of drugs. The main rule, however, is that persons without competence to consent should only be involved in research when the study cannot be performed on people who are able to give their own consent, and when it is likely that the research is of direct and significant benefit to the individual or group participating in the research.
- Consent to research involving children and adolescents must be given by parents or persons with parental responsibility. Children who can express their own views in the case then have the right to be heard before the guardian makes a decision, and the higher the age and the greater the maturity of the child, the greater the emphasis on the child's opinion must be. Under special conditions, adolescents themselves can consent to certain types of research.
- Each country has detailed regulations for this, which must be followed.

• Consent must be given *freely*.
Anyone who is asked to participate in a research project must:

- Be informed that participation is voluntary.
- Have a real choice between participating and not participating.
- Not be pressured to participate. The threshold for what can be defined as 'pressure' is low. An example of pressure might be to say that 'everyone else we have asked wants to participate'.
- Not be tempted to participate, for example in the event of an offer of improper payment or other benefits (compensation for expenses for participation and any normal fee for the time spent will normally be acceptable; see Sect. 13.13).
- Not be asked for consent from anyone close (a colleague, manager, family, friend or the like).

When obtaining consent, experience has shown that the condition of the persons who are to give their consent, and the situation they are in, sometimes makes it doubtful whether consent is given freely, even if one follows the guidelines above. The problem is particularly relevant in research involving, for example, prison inmates, drug addicts, children, institutional residents, the homeless, the terminally ill, refugees and asylum seekers, etc. (often the term 'vulnerable groups' is used as a somewhat vague collective term for these). Some examples can illustrate the problem:

Examples: An asylum seeker may, in the unfamiliar and frightening situation she is in, mistakenly believe that participation in the research may in one way or another be positive for the asylum application. A child may respond obediently to what an adult asks. A dying person may say yes in pure apathy. In these cases, the consent is not given freely in the true sense of the word.

When obtaining consent from people who belong to particularly vulnerable groups, one must therefore often make an additional effort to ensure that the consent is truly given freely. This requires competence about these groups and an ability to understand the person from whom one is seeking consent.

- The consent must be *documented*.

  - This requirement can be met in several ways:

    ▪ In writing.
    ▪ Electronic, provided the person's identity is verified electronically ('digital consent').
    ▪ Oral, documented with audio/video recordings or similar (this is only relevant in special cases, for example when asking for consent from illiterates).

13.6   Special Measures for Involving People Who Do Not Have Competence to Consent

When a person does not have the capacity or competence to give consent, others (a relative, guardian, legal representative, etc.) may, under certain conditions, do so on his or her behalf. It goes without saying that this should be severely restricted. National laws and institutional regulations will prescribe the circumstances for this, but on a research ethical basis the prerequisites should be:

- The person involved must not object to participating in the research. For children, this applies from the age the individual is able to express such reluctance.
- The potential risk and burden on the person involved must be negligible.
- There is a real reason to assume that the research results may benefit the person involved, or benefit other persons in the same category.
- Corresponding research cannot be carried out with persons who have their own competence to consent.
- There is reason to believe that the person in question would not have objected to participating in the research if he or she had had the capacity to do so.

It is important to note that any consent from relatives, guardians, etc. must be set aside if the person in question opposes participation in the research. This also applies if it happens during the research. In addition, it may be irresponsible to initiate research without sufficient professional competence, and preferably experience, on the relevant category of people.

The requirement that a relative, guardian, etc. must give consent when the person involved does not have the capacity or competence to do so, may under certain circumstances be waived in the event of a clinical emergency where there is no time or it is practically impossible to obtain consent.

13.7 Right to Withdraw from Further Participation in Research

Everyone who has given consent to be involved in research has the right to withdraw from further participation at any time without giving a reason. Information about this right must be clearly stated in the consent declaration signed by the participant.

13.8 Special Requirements for Initiating Research Involving Humans *Without* Obtaining Consent

In exceptional cases, prior information and consent from the participant are not required. Within medical and health research, this is regulated in national legislation and limited to clinical emergencies. In other research fields, there are further possibilities for which some examples may be illustrative:

Example 1: For methodological reasons, in some research projects it may be necessary to observe people's behaviour without them being aware that they are being observed. In other projects, there may be methodological reasons to inform the participants in advance but without giving them the full or actual purpose of the research. This could be acceptable when there are no alternative procedures, when the research is important and the foreseeable burdens for the participants are small. In all cases, the participants must afterwards receive full information about the project, what they have participated in, and why they were not informed in advance. Furthermore, the information obtained about the participants cannot be used without written consent from them (in accordance with the fact that everyone has the right to withdraw at any time and for no reason)

Example 2: In some research projects, people may be indirectly involved in the research. It is, for example, conceivable that research on the situation of inmates in a prison could generate information of a confidential and perhaps sensitive nature both about other inmates (who are not included in the study) and about employees in the prison. The same may happen in studies of groups within a company or organization. Here it may be acceptable to begin the study without obtaining the consent of everyone who could possibly become indirectly involved. In such cases, information that has accidentally emerged about people who have not given their consent cannot be used without their subsequent consent. With regard to personally identifiable information, special rules apply here (see Sects. 13.5 and 13.9)

Example 3: Some research projects involve observing or otherwise studying people who are essentially anonymous, without being in direct contact with them. It may be possible to do this without obtaining consent if the research is sufficiently substantiated. However, if the observations are recorded with images, video, audio recorders, etc., one should inform persons in advance, and if the study accidentally leads to personal information about identified or identifiable persons, further processing of this information will require prior consent (see Sects. 13.5 and 13.9)

The requirement to obtain consent from those involved in research is in some cases an obstacle to research. Some, therefore, believe that the legal provisions are too rigid, or that they are interpreted and practised too strictly. Particularly in medical and health research, some believe, for example, that the strict consent requirement in certain cases acts as a bureaucratic obstacle to research that could lead to better treatment for sick people. Others view this differently. The consent requirement and the way it is practised are therefore frequent topics of debate. Whenever there is room for discretionary interpretation of the rules on consent, returning to the underlying ethical principles of protecting and respecting people can often help to clarify matters.

13.9   Statutory Requirements for the Processing of Personal Data

Personal information obtained from registers, archives, interviews, surveys, the internet, etc. is a highly important prerequisite for research in many fields. At the same time, this is information every individual to a great extent has the right to decide over for themselves. The processing of personal data is, therefore, both covered in research ethics guidelines and carefully regulated in national laws to ensure that the rights, integrity and privacy of individuals are respected and protected. An overview of the legislation in different countries can be found in an online handbook published by a global law firm (DLA Piper, n.d.). In Europe, the national legislations in this area are harmonized via the EU's General Data Protection Regulation (GDPR) (European Union, 2016), which came into force in 2018. These laws do not apply specifically to research activities, but are important for researchers because archives, registers, etc. are central sources for research in many subject areas. They have detailed provisions which all researchers who are to process personal data must follow.

Research organizations that process personal data are required by law to have detailed procedures for this which the researcher must follow (see later). Nevertheless, in this book it may be appropriate to summarize the main points. The review is based on European legislation, but can also be informative for researchers in other countries. However, the reader should keep in mind that terms, definitions and detailed provisions differ from country to country.

- The term *'personal data'* means information relating to an identified or identifiable living individual in the form of numbers, text, images, audio/video recordings, etc. This will include:

  - Information that can be linked *directly* to a person, for example in that their name, identification number, address, email address, personal characteristics, photos, videos, audio recordings, etc. are part of the information.
  - Information that can be linked *indirectly* to a person, in that one is able to uncover who the person is from the content of the information, even if no name, identification number or the like is given. This also includes information that can only be linked to a specific person through a more elaborate investigation into the information or information that makes it possible to narrow the identification to someone in a small group of people (the latter can form the basis for speculation that can be harmful to those concerned).
  - To prevent the information from being linked to individuals, it is common to *pseudonymize* it, i.e. remove all information that identifies the person, but link a pseudonym, a registration number or similar to the other information about the person. The pseudonyms are then generally linked to the natural persons by a list of names that is kept secret. For the pseudonymized information, one must also ensure that there are no combinations of sub-information that can identify the person or narrow the identification to a smaller group. Both the process of pseudonymising personal data and further research on these are covered by the law.
  - Research on information about completely *anonymous* people (no one holds a key that can link the information to individuals) is not covered by the law.
  - Research on *deceased* people is mainly not covered by the laws regarding handling personal data, but research ethics issues are related to such research as discussed in Sect. 13.15.
  - For *special categories of sensitive personal data*, separate provisions apply. In the European GDPR, this includes information about:

    - Racial or ethnic origin, political opinions, religious or philosophical beliefs.
    - Genetic and biometric data when the purpose is to uniquely identify a natural person.
    - Health.
    - Sex life or sex orientation.
    - Membership in trade unions.
      As a general rule, the processing of such information is prohibited, but there are exceptions.

  - The processing of information that a person has been suspected, accused or convicted of a *criminal offences* can subject to the European GDPR only be carried out under the control of an official authority. Whether and how such information can in practice be used for research purposes must be clarified on a case-by-case basis.

- Information on a person's *health* is subject to both the legislation concerning processing of personal data and special legislation related to the use of health data.

• The term *'processing'* of personal data means the collection, registration, compilation, storage, disclosure, deletion and more – or combinations thereof – and applies to:

  - Processing of personal data that is carried out in whole or in part in an automated manner.
  - Non-automated processing of personal data that is included in or is to be included in a register (registers, lists, etc. where personal data is stored systematically so that information about the individual can be found).

• When personal data are to be processed in connection with research, one must in most cases have the *consent* of each person to whom the data applies.[1] Specified requirements are set for the form and content of a legally valid, informed consent (see Sect. 13.5). The duties of the researchers, which will constitute the precondition for consent, must be clearly expressed. This must include a description of how the information will be treated confidentially, and how pseudonymisation and other measures will prevent the information from being linked to the person giving consent (for a promise of confidentiality and its limitations; see Sect. 13.12).

  In a research context, it is particularly important that consent can be given quite broadly, within certain research areas, provided it is within recognized research ethics norms. The data can then, under certain conditions, also be processed further in new projects without renewed consent, provided that the person responsible for processing the data is the same.

• If there is a high risk that a planned processing of personal data may violate a person's rights and freedoms, the person responsible for the processing must, before the project starts, make an assessment of the impact the processing may have on the person concerned (Data Protection Impact Assessment, DPIA). The assessment must follow special procedures.

National legislation also sets out detailed requirements for organizations where personal data is processed. The purpose is to ensure and document that all employees follow the legislation when they process personal data. The legislation requires each organization to establish internal procedures for processing personal data. This includes the appointment of one or more persons with overall responsibility to ensure that all employees comply with the law. They must also establish an infrastructure for secure storage of personal data, for archiving and documentation of the

---

[1] *Consent* is one of several alternative grounds for processing personal data legally according to the European GDPR. Some of the others are irrelevant to research. Some may be relevant in certain research works, but require relatively comprehensive assessments and justifications. In such cases, consent may be an easier option.

personal data to be processed, for quality assurance, system evaluation, and more. These legal requirements vary somewhat from country to country, and each organization has some freedom to organize itself in the best possible way in relation to its own needs.

In practice, therefore, a researcher who is planning to collect and use personal data in the research can simply follow the details of the research organization's internal procedure step by step and end up with a project plan where individuals' rights and freedom are respected in accordance with the law. The organization's administrative resources should also be able to provide guidance on unusual ethical and law-related issues that may arise.

13.10   Requirements for Prior Approval of Biomedical and Behavioural Research Involving Humans

In most countries, some types of medical, health and behavioural research involving humans must be *approved in advance by an ethics review board* (ERB) (also called institutional review board (IRB) or given a related name, depending on country). Depending on the national regulations, the boards can be national, regional, institutional, or other groups of impartial and experienced persons. The board's responsibility is to safeguard the interests of humans involved in research and to protect them from risk and burden. It functions partly as an element in the quality assurance of research proposals focusing on ethical and legal requirements, partly as an authority that can accept or reject a project proposal on ethical or legal grounds. The types of research that need prior approval, and the organization, mandate, appointment and operation of ERBs are regulated in national laws and the research organizations' internal procedures. Readers who for some reason need information on the regulations in foreign countries may start with an overview provided by the Office for Human Research Protections (n.d.) (a US governmental agency). ERBs are required especially in medical health-related research; see CIOMS (2016, guideline 23).

An approval from an ERB does not release the researchers from their ethical and legal responsibilities, but will naturally be important for the assessment of guilt if something goes wrong.

13.11   Ethical Issues in Research on People on the Internet

13.11.1   The Internet – A New Research Area and a New Source of Information

The widespread use of the internet by the general public has opened up many new research topics and new research methods, so-called 'internet research'. This spans many research fields, but only internet research involving people is discussed in this chapter, i.e. research where information and data about people, their opinions, actions, use of and behaviour on the internet, etc. are retrieved from the internet.

The internet is flooded with individuals' utterances about almost everything – opinions of a political or ideological nature, personal thoughts and intimate confidences, descriptions and pictures from travels and gatherings, communication between friends and people with shared interests, etc. This constitutes a rich source of information for research in many subject areas. Internet research is therefore an increasing research arena. However, obtaining information on individuals from the internet also raises a number of new ethical, methodological and practical issues that require special care and consideration. The research community's experience with these issues is still limited, although a number of ethical guidelines for internet research have been established. Some of the issues that have been discussed in research ethics literature and mentioned in guidelines for internet research in recent years, for instance by Fossheim and Ingierd (2015), NESH (2019), and Franzke et al. (2020), are summarized in the following.

13.11.2 Guidelines for Internet Research Involving Humans

Internet research involving humans is in principle subject to the same ethical guidelines as other research involving humans. This implies, for example, that:

- The researchers must make themselves known to and obtain informed consent from those they plan to observe or involve in research on the internet – with certain exceptions.
- The researchers must process personal data they obtain from the internet in accordance with the law, as described in Sect. 13.9.

This applies whether the information is available to everyone or subject to limited access on the internet. The problem is that these two requirements can present practical difficulties:

13.11.3 Uncertainty About What Is 'Private' and 'Public' on the Internet

One of the most basic ethical requirements for research involving humans is respect for the *privacy* of the individual. One aspect of this is respecting that people want to keep some of their own thoughts and opinions and some of the information about themselves secret, or only to share it with others they trust. Everyone has a *private sphere*, where certain things are only available to a select few, and a *public sphere*, where one is willing to express some of one's thoughts and accepts, or even actively seeks, to make information about oneself known. Having respect for people's privacy in research is largely about understanding what the individual's choices entails, and carrying out the research so that the individual's private sphere does not become (more) open to the public *as a result of the research*, without prior informed consent to do so.

This may often give rise to difficulties in internet research because many of those who comment or post information on themselves on the internet is unclear about what is 'private' and 'public' and may have insufficient understanding of this distinction. *Friends*, for example, are traditionally a small group of people with whom one regularly associates and regards as part of one's private sphere. On the internet, you can have a thousand 'friends' who you have never met face to face and who you usually know little about. As in ordinary friendships, 'friends' on social media may share opinions and interests and feel an intimate closeness to each other (which can perhaps be compared to old-fashioned penfriends). In reality, they may be complete strangers to each other. But in contrast to traditional meetings with friends within the private sphere, meetings with friends in social media often takes place in a public setting, leading to confusion about the boundaries of the individual's private sphere. For example, a blogger may want to share personal or perhaps intimate reflections and experiences with others – perhaps to help people in the same situation or to get support from others who read the blog. At the same time, the blogger may also want readers to treat this respectfully, as something confidential, which in reality is wishful thinking, no matter whether the statement takes place in public on the internet or in a more closed online forum. Furthermore, the fact that people talk openly about something on the internet cannot be taken as an indication that they agree to let it be subject to research. To get or give help, for example, many people write on the internet about their own illnesses and health, even in open online forums. For many, getting information on this noticed, interpreted and commented on in a scientific article may be a far greater exposure than they were initially willing to subject themselves to. It must therefore be assumed that many of those who post information on the internet do not want it to be subject to research, and do not find it beneficial for them that the information becomes more accessible and more widely known than it already is.

In respect of people's privacy, research ethics guidelines for internet research therefore emphasize that researchers who want to use statements from and information about people published on the internet should in principle consider factors such as:

- How accessible and public the statements and information actually are (whether the website is open or closed, the degree of any age and access restriction, the number of people that have access to or visit the website, etc.).
- How public the information is *intended to be*.
- The extent to which those who express themselves and post information are able to understand and take responsibility for what they do.
- The context in which the statements and information are posted on the internet.
- How sensitive the information is.

On the basis of on an overall assessment of such factors, one must then form an opinion as to whether it is necessary to ask for consent to use the information in research. The reader may find more detailed advice on this in national or institutional guidelines.

In addition, one must also consider whether the planned research will include the processing of personal data in a way covered by national legal provisions as described in Sect. 13.9. In that case, informed consent is usually needed, regardless of the ethical assessment above. Obtaining consent can, however, often present major practical problems; see the next section.

13.11.4   *Problems Communicating with and Obtaining Consent from People Active on the Internet*

With few exceptions, involving people in research requires communication with them: They must be contacted, told who the researchers are and be informed about the project; give consent to participate; be kept up to date during the work and informed about the results, etc. This can present great difficulties when those involved only appear on the internet:

- For an informed consent to be valid it must be given by a person with a known identity. On the internet, however, there are some who act under a pseudonym or who simply hide under a false identity or pretend to be someone other than the person they really are. In such cases, obtaining consent from people with a reliable identity can be a practical problem.
- Postings and information accumulated on the internet may have been there for a long time. This can be used in research, but finding the people behind it and obtaining their consent can be very difficult. This is also a scientific problem because the accuracy and credibility of old internet information can be uncertain.
- In internet research, the persons subject to research can have communicated on the internet prior to the researchers' arrival. They can also communicate with each other independently of the researchers while the research is in progress. This makes it difficult for researchers to control the content and flow of information about the research project and to ensure that everyone perceives the information equally and correctly (from time to time, similar issues can of course also arise in research on groups outside the internet).
- Those who are active on the internet 'come and go'. Researchers who want to observe or intervene in an internet forum or similar may then have difficulty keeping track of who is involved, obtaining consent from new members and providing equal information to all.

13.11.5   The Risk that Anonymous or Anonymized Information from the Internet May Be Traced to Its Source

There are occasions when researchers may be interested in statements, information, etc. posted on the internet anonymously, or that deal with unnamed persons. Essentially, this can be used in research without anyone's consent. However, *when the research results are published*, anonymous material retrieved from the internet can often still be traced back to the website the material comes from via internet search engines, especially if the researchers quote verbatim from the internet text or use many of the same keywords that were used in the original text (a well-considered rewriting of the original texts can reduce the possibility of this). From the original, anonymous internet page, people with the right skills and opportunities can then in many cases find their way back to the source, a named person, via electronic tracing. At other times the information may be so specific that the anonymous person who posted it can be identified, at least by family, friends and colleagues. In both cases, the information cannot be regarded as anonymous information, but must be considered as potentially personal data that may have to be handled in accordance with national law. This means that researchers who use 'anonymous' material from the internet must exercise great care and critically assess whether the information can in some way be traced or linked to someone so that consent is needed.

13.12   Pledge of Confidentiality and Its Limitations

13.12.1   Pledge of Confidentiality – An Important Element in Safeguarding the Interests of People Participating in Research

Another important element in protecting human rights and respecting the dignity and privacy of people who are invited to participate in research (in addition to asking for their consent) is to *promise* them that data and information obtained from or about them will be treated *confidentially*. In this respect, confidentiality is looked upon as a duty in research ethics. In addition to safeguarding the interests of the participants, a promise of confidentiality is also often necessary to get people to participate in research projects. In practice, the pledge of confidentiality is commonly formulated as one of the preconditions for the written consent from the participants. At the same time, the researcher must always also inform about what the pledge of confidentiality entails. This includes information on:

- How information and data in practice will be collected, stored and possibly destroyed; who gets access to the material; what one does to prevent others from accessing it; how the participant's anonymity is ensured in project reports and scientific publications, etc.

• The limits for how far a promise of confidentiality applies (see section below). This information may be vitally important for the informants.

In certain contexts, a promise of confidentiality is also *required by law*. The scope of this statutory duty in connection with medical and health research and processing of personal data is mentioned above, but other laws may also contain provisions on the duty of confidentiality. The readers must familiarize themselves with the laws that apply in their own country. The research organizations will also have internal procedures for this, both in terms of what is to be treated confidentially and how confidential material must be processed.

13.12.2   The Limitations of the Pledge of Confidentiality

Research involving people occasionally reveals conditions or situations to which no one can or should be neutral and passive. Examples are information that a person has committed a serious crime, or that someone who has been convicted of a crime is actually innocent; that someone is being abused or mistreated at home or in the workplace; about sexual abuse on the internet; on serious medical malpractice; that someone is planning terrorism or otherwise poses a danger to society, etc. Everyone who becomes aware of this has a *moral responsibility* to react to *prevent* illegal acts, injustice and abuse from taking place. The majority of people would probably also find it important and morally right to contribute to criminals being held accountable for their wrongdoing. However, if information on matters like this emerges under the promise of confidentiality, the researcher is faced with an ethical dilemma where two ethical principles must be weighed against each other. On one hand, adhering to the promise of confidentiality, i.e. respecting the interests of the informant; on the other, trying to prevent anyone from being seriously harmed, i.e. breaking the pledge of confidentiality by reporting to relevant authorities. It goes without saying that the latter option is only relevant to consider in serious situations. In some cases, though, researchers may find a way to avert a criminal act without violating their promise of confidentiality by reporting to the authorities.

In some contexts, the ethical dilemma related to keeping or breaking a pledge of confidentiality is clarified by national law. The provisions will vary from country to country and are often elaborate. They typically include statutory obligations to:

• *Report to the appropriate authority if one becomes aware that a serious criminal act may be committed, provided that the reporting can prevent the crime.* The duty may also apply if serious consequences of a criminal offence already committed can be averted. This duty to prevent may apply even if the information has been obtained confidentially. Statutory duties to report and avert apply only to serious crimes such as murder and grievous bodily harm, rape and incest, hijacking, matters related to national security, etc. Reporting of information that can help to solve criminal offences or bring them to light are generally not statutory. These legal provisions essentially apply to everyone, including researchers, but priests, lawyers, doctors and others are subject to special provisions that exempt them from disclosing certain types of confidential information.

- *Report to appropriate authorities on any reasonable suspicions of child abuse and neglect.* This duty may apply to everyone, including researchers, other professionals and officials such as hospital and health care employees, teachers, police officers, etc. even when the information is otherwise confidential or privileged. This reporting duty is a legal requirement in many, but not all, counties. In the absence of legislation, public regulatory bodies, professional associations etc. may have code of conduct that requires certain professionals to report suspicions of child abuse and neglect.
- *Testify as a witness in criminal cases.* This may include revealing research data and information obtained under confidentiality. The legislation on this is usually detailed and complex.

In addition to this, there may be national laws that oblige researchers, in other contexts than those specified in the points above, to disclose sources of information and confidential information obtained through research.

Researchers who issue a pledge of confidentiality to someone have a duty to inform about the statutory obligations they have in certain situations to report and testify about confidential information. The informants may then choose not to disclose information to the researchers that could harm them if reported to the authorities. However, these issues are somewhat up for debate in the research community. For instance, some argue that this may reduce the reliability of research results and make certain types of research difficult to conduct. Others believe that this problem can be prevented to a great extent through carefully designed research methods. Some believe that a legal requirement that researchers must, under certain conditions, provide confidential information to the authorities, cannot be defended ethically, i.e. that researchers should oppose the law and rather take the consequences of breaking it or not obeying a court order – ethics must take precedence over the law, in other words. However, the laws in this area are rooted in common morality and justified to avert crimes and prevent people from being harmed. The order of the law must therefore be seen as an expression that society finds it ethically more important to prevent people, and especially children, from being harmed than to respect an informant's right to keep something confidential. So, the primary question here is not about ethics versus the law but about weighing conflicting ethical principles against each other.

Some researchers argue that a pledge of confidentiality made by a researcher should have as strong legal protection as a pledge given by, for example, a lawyer or a priest. However, there is a difference here: The lawyer's promise of confidentiality given to a client is strongly protected so that the lawyer can give the client the best possible defence and legal help. The priest's promise of confidentiality is strongly protected because the priest can help give people peace of mind. Researchers normally have nothing to 'offer', there and then, that can be of direct benefit to those who are promised confidentiality.

Knowledge and experience in handling situations where a pledge of confidentiality conflicts with a duty to provide information to authorities should be viewed as part of the professional expertise of researchers working in disciplines where this may occur. It should therefore be a natural part of the researcher education within the subject.

13.13   Payment for Participation in Research

People who agree to participating in research, whether it is answering a questionnaire, being interviewed, or being the subjects of experiments and clinical trials, make their own time and energy available to the researchers and are willing to expose themselves to discomfort and inconvenience that may be associated with their participation (sometimes repeatedly and over a longer period of time). It is not unnatural to see this as a small part-time job one should be paid for. Participation may in some cases also lead to lost earnings and expenses, for example for travel, board and lodging, which should also be compensated. The latter is unproblematic, but paying a fee or giving another form of compensation for the participation itself raises a number of methodological and ethical issues about which there are a range of opinions in the research community. Since this is most often a question of participating in non-profit and non-commercial research activities, most people will ideally want the participants to contribute for free. Many potential participants also see it this way, at least when it comes to certain types of projects and when the usefulness of the research is well explained.

However, on some projects, researchers sometimes find it difficult to recruit the required number or the preferred selection of participants unless a payment or other form of direct compensation is offered. In practice, it is often a question of a smaller amount, a lottery ticket or the like. However, some believe that this is to 'entice' people to participate – and that this is ethically wrong and methodically problematic:

- If the compensation is set so high that in practice it acts as a lure, one can ask whether the consent to participate is as voluntary as law and ethics presuppose. For example, it is conceivable that some people will accept a higher risk than they ought to, just to make some money. This possibility requires the exercise of great caution in determining the level of compensation.
- In many research projects, it is important that the selection of research participants is right in relation to the issues of the research. If the compensation is used as bait, the composition of participants may become skewed. One can imagine, for example, that a poor person is more easily lured by compensation than a rich person. However, some point out that those who are difficult to recruit to participate in research are often also the ones who respond most positively to compensation. They therefore believe that the compensation can be scientifically justified because it may contribute to a more representative selection of research participants.

- Some also point out that compensation for participation may affect the participants' answers in an interview, questionnaire or similar.

There can be several reasons why people do not want to participate in a research project. One reason may be that they perceive the project as so unimportant, strange or distant from their own reality that they have no interest in participating. When people are negative towards participating in a research project, it should therefore lead the researchers to look critically at their undertaking. Is it really important enough? Has its importance been well explained?

Payment of research participants has been discussed for the most part in relation to medical and health research, and there may be relevant national and institutional guidelines for compensating research participants in these fields. Such guidelines may also be relevant in other research areas. As an example, some of the principles in the guidelines from The [Norwegian] National Committee for Medical and Health Research Ethics (NEM) (2009) can be summarized as follows:

- Those who participate in research are entitled to coverage of direct expenses and a possible loss of earned income in connection with the participation.
- Those who participate in research should in principle be invited to contribute to the research free of charge. Alternatively, a reasonable gift may be given in recognition of their participation.
- Payment for time spent and any inconveniences or burdens are, however, acceptable and will in certain projects be both fair and reasonable. It is also acceptable to use payment as a tool to obtain participants, although this should preferably be avoided. However:
  - The payment must be linked to what the participant actually contributes, and must be in proportion to the time spent and the extent of any inconveniences and burdens. Payment does not legitimize increased risk; in all cases the risk must be low.
  - Payment must be reasonable/moderate.
  - Payment should, if possible, be standardized (fixed hourly rates, unit price for interviews, etc.).
  - Participants who withdraw along the way must normally be paid for their participation until they withdraw (the terms of payment must affect a participant who wishes to withdraw as little as possible).
  - The payment can be replaced by other forms of compensation (tickets in a lottery, a gift item, etc.).
  - There must be transparency about the payment (openly described in project plans, project reports and more).

13.14 Consideration for Those Who Are Not Directly Involved in Research, But Who May Be Affected by It

Research involving people does not only affect those who are directly involved and have given their consent to participate. Family, friends, work colleagues and other close relations can also be exposed to burden and, in the worst case, be harmed by research that they are only indirectly affected by. Occasionally others can also be exposed to burden and harm.

In many cases, those who participate in a research project may be more tolerant of the burdens of participation than close relations and others who do not participate directly. In a research project on drug abuse, for example, one can imagine that a drug addict who is interviewed does not feel particularly stressed about opening up to a researcher, while a spouse or cohabitant and children may feel this as an unpleasant and frightening revelation of family privacy, even if the study in no way applies to the family relationship. Burden and harm can also be caused by worry and fear. It cannot be taken for granted that the interviewee understands this when consent to participate is given.

Any burden and harm to close relations and others often manifests itself when the research work becomes publicly available and contains information of a negative nature for individuals, groups or organizations. Everyone finds it stressful when issues one somehow feels connected to are highlighted, problematized, analysed and discussed in public, and for particularly vulnerable people, attention in itself can feel like abuse.

Although legislation and research ethics focus on safeguarding the rights and welfare of people who directly participate in research, it is no less important that researchers do their best to ensure that others who for one reason or another may be indirectly seriously harmed by a research project, are also safeguarded. Research projects involving humans should therefore be planned so that the burden on everyone who may be indirectly affected by the work and the results is also minimized. Examples of responsible practice in this context could be to:

- Inform relatives and others, who may be indirectly affected, about the project before it starts. Listen to their opinions and possible concerns and take this into account in the planning and implementation of the project. If the usefulness and importance of the work do not clearly outweigh the burden on participants, relatives and others who may be indirectly affected by the project, it should be changed or not initiated.
- Formulate the research report in such a way that close relatives and others who may be indirectly affected by the work being published are not exposed to unnecessary burdens.
- Be thorough with anonymization.
- Be careful not to draw conclusions or generalize beyond the facts.

Special cautions must be taken when children may be indirectly affected, or when the research concerns small, transparent and vulnerable groups of people.

13.15   Respect for Dead People

13.15.1   *Respect for the Deceased's Reputation*

Research involving dead people is central to some disciplines. Such investigations may include information and stories about the deceased that are found in private and public archives, registers, documents, books, etc. Dead people can also indirectly contribute as an 'informant' in the research through written material, audio recordings, photos, films and videos that they leave behind. In certain disciplines, living people are also an important source of knowledge about the deceased. In such cases the question is what ethical and possibly statutory responsibility one has when deceased people are the subject of research.

The legislation that protects a person's dignity, human rights and privacy, applies with a few exceptions only to *living* people. An example of an exception is health information about the deceased. Another is personal data about a deceased person that may reveal something about identifiable living persons (for example, information about hereditary diseases). Processing of such information then requires the consent of the living persons affected. Some countries may have legislative provisions specifically relevant to research involving deceased persons. In the absence of legal regulations, however, ethical considerations must be taken into account to distinguish right from wrong when processing information about deceased persons and writing about them. Two issues then appear to be particularly relevant to consider:

- The question of whether a living person has 'rights' and 'interests' that persist and possibly develop further even after that person's death, what these are in that case, and whether they can be violated even after death.

  - Since ancient times, philosophers and others have pondered whether it is at all possible to do injustice to dead people. One of the arguments for the impossibility of this is that dead people are just corpses that are unable to feel injustice, burden or pain. Similarly, it is also unreasonable to talk about a corpse having human dignity, human rights and privacy. The ethical justification for treating living humans with respect can therefore not simply be used as a justification for treating dead people with respect.
  - Nevertheless, there are both legal rules and general acceptance that living people in certain cases can make decisions that others must respect after the person in question has died. This primarily includes decisions related to the transition from living to dead. In life, one may for example:

    - Draw up a will that determines how one's assets and wealth are to be distributed after death.
    - Decide that one's remains may or may not be used for organ donation or for research.
    - Express a desire about how one's funeral should be and how one's remains should be taken care of.

- Express a desire about how personal documents, memorabilia, etc. should be disposed of (destroyed, given to collections, etc.).
  Such decisions or expressed desires are generally respected and complied with as far as is possible within the bounds of the law. So, these are examples of rights and interests people have in life which are commonly respected and upheld after death.

  – Many people probably see it as an important criterion for a good society – where people have empathy and concern for each other – that details about one's own private life (which one chooses to keep for oneself) should not be revealed, unfolded, discussed or researched by others after death. This can, for example, include matters related to views on life, health, life and work with others, sexual relations and wrongdoing one regrets. A society where both common morality and the researcher's morality require respect and care for dead people is therefore of great general interest and value to many living people.

- The question of information about dead people, and the stories told about them, can be harmful or burdensome to relatives, descendants or other living people or groups of people.

  – When living people are involved in research, family and others can also be indirectly exposed in various ways to burden and, in the worst case, harm, as addressed in Sect. 13.14. Similarly, it is easy to imagine that research involving *dead* people can be stressful and, in the worst case, harmful to *living* people who are in some way connected to the dead.
  – Within some cultures and religions, dead people have a special position or are the subject of particular respect. Actions that violate the notion of dead people, or that are otherwise perceived as disrespectful, can be a burden and harm to people belonging to such cultures and religions.

The diversity and complexity of these issues means that researchers often have to use their own ethical and professional discretion to draw the line between what is respectful and responsible and what is not. In practice, therefore, researchers may think differently here.

Example: A historian writes the story of a company that in its time was founded and built up by one of the country's most famous and renowned industrialists – a charming but authoritative personality. Ten years before his death, the industrial entrepreneur had published an autobiography about his life, starting as a poor child of the underclass and ending as a rich and respected member of society. Because the company and the entrepreneur are so closely linked, the historian believes that the entrepreneur must also be mentioned in the company's history, but now in light of the company's development. With his autobiography as a starting point, he provides new information through interviews with employees and others who knew the entrepreneur. The family also gives him access to a number of storage boxes in the attic, with memorabilia and documents that span most of the industrialist's life. In one of these, the historian discovers to his surprise that some papers from the industrialist's youth, when the country was occupied by Nazis, might indicate membership in the youth organization of a nationalist party collaborating with the occupants – or at least participation in a couple of sports championships they had arranged at the very end of the war.

From other sources, he gets the participation in sports championships confirmed but finds nothing about membership. In the autobiography, the interest in sports had been mentioned with obvious pleasure, but the participation in Nazi-organized sports was not. In adulthood, the industrialist had never said or done anything that might indicate any Nazi ideas or sympathies. The historian nevertheless finds the new information important in understanding the industrialist's background and personal development and that it provides an interesting picture of the time he grew up. In the book, therefore, he chooses to discuss and comment on the facts he has found. When the book is published, the information about participation in Nazi-organized sports activities is given great attention in the media. A newspaper fills the front page with an archive photo of the industrialist in white tie, on his way to a dinner at the royal castle on the occasion of a state visit, with the headline: '[The name] participated in Nazi-organized sports'. The story of the company is of little interest to the media, but the revelations from the industrialist's youth increase the sale of the book. Both the company's management and employee representatives, and the family, react to the fact that the information about the young boy's participation in Nazi-organized sports is included. They point out that his sports activities cannot be used as indication that he had Nazi sympathies in his teens, and that nothing in his later life suggested such sympathies. They therefore find the information irrelevant to the company's history and take the view that it does not change anything about the founder's ideals and deeds. They also point out that other, obviously relevant information about the industrialist has not been included. They believe the historian had to understand that his choice of material and formulations would provide a basis for unbalanced speculation and gossip that would not only undermine the industrialist's legacy in an unfair way but also indirectly damage the family and company's reputation. A suspicion was sown which no one could completely remove. They also point out that an annual prize that the industrialist has donated to young researchers will no longer feel as rewarding to receive. The critics in this case are supported by other historians, who believe the author's selection of material is worthy of criticism both ethically and scientifically. Others support the author.

In this example, it is indisputable that the historian triggered media coverage that both weakened the industrialist's reputation and burdened and harmed others. The question is, however, whether the new information about the industrialist has such great significance for the history of the company and its founder that its negative consequences can still be justified. Here, it can be easy for a researcher to underestimate the seriousness of the damage inflicted on the deceased's legacy, and the burden other people are exposed to, at the same time as the significance of their own findings is overestimated. However, when showing 'respect' for the deceased and others affected, it must be expected that what may have been the deceased's opinion, and what is the opinion of the living affected, is given special weight.

In this example, it is also indisputable that the historian's way of discussing his findings led to sensational headlines in the newspapers that gave a skewed picture of the industrialist. In the context of research ethics, the question then becomes whether the historian showed sufficient caution towards the consequences of his actions. Most people will probably agree that in this case he should at least have mentioned and commented on the facts in a way that could have reduced the possibility that the new information about the industrialist could unfairly damage his legacy and be misused in sensational headlines.

13.15.2   Respect for Graves, Skeletons and Other Human Remains

Graves and human remains are objects of respect in all countries and within certain religions and cultures have a particular value and cultural function for descendants who identify with the deceased. There are therefore both national laws and special ethical guidelines for handling graves and burial materials. The regulations will vary from country to country, but the thinking is the same. As an illustration, a few guidelines for the ethically responsible handling of human remains are provided below (again based on a Norwegian guide prepared by the National Committee for Research Ethics on Human Remains (2019)):

- Respect for the deceased, i.e.:

  - Ensure that human remains and burial sites are treated with dignity and discretion.
  - Clarify and take into account what the deceased's own wishes and opinions were or probably might have been, as far as it is possible to know or imagine. The more recent the remains, the more important this is.
  - Be equally respectful of the remains of people of all kinds.
  - Be especially cautious when there is uncertainty about the origin of the remains, for example, clarify whether remains stored in collections may have resulted from human rights violations, crimes, illegal trade in remains and more.

- Respect for descendants, i.e.:

  - Clarify and take into account the wishes and opinions of any descendants. The more recent the remains and the closer the relationship is, the more important this becomes.

- Respect for groups of people who affiliate with the deceased (culturally, religiously, ethnically, nationally, etc.), i.e.:

  - Clarify and take into account the wishes and opinions of groups of people affiliated with the deceased or who identify with the deceased, whenever this is possible and relevant. This is especially important when the group has distinctive views on death and dead people.

- Respect for the rarity of human remains, i.e.:

  - Avoid destruction of rare and unique material.
  - Facilitate that the material can also be used by other researchers.

- Assessment of the project's quality, feasibility and importance in relation to its negative consequences, i.e.:

  - Do not initiate projects where the quality and the feasibility is uncertain, unless there are special reasons to proceed.

 – Assess any negative consequences of the project and only start the project if
   its scientific importance or societal benefit clearly outweighs the negative
   consequences. Negative consequences can include, among other things, that
   the deceased's legacy risks significant damage, or that descendants and others
   affected are exposed to burdens or harm in various ways. A negative consequence is also that rare material is destroyed as a result of the project.

• Respect for relevant laws and regulations.

This applies to both expected and unforeseen discoveries of human remains.

The guidelines for the handling of human remains often require a good deal of
ethical discretion in order to comply with them in a responsible manner. In particular, this discretion will be put to the test using modern methods of analysis (for
example analyses of genetic material) that can shed completely new light on the
identity of kinship, lineage, living conditions, diseases, physical and mental characteristics, etc. Such analyses can penetrate deep into the private spheres of the
deceased and perhaps reveal things they would rather have kept to themselves. On
the other hand, the surveys can help to draw a more accurate and detailed picture of
the deceased and the history of which they were a part. Weighing negative and positive effects against each other is difficult, and the result of such discretionary assessments is expected to vary from researcher to researcher. As such, this favours
seeking advice from experienced colleagues and relevant ethical committees.

13.16   Respect for People with Values One Does Not Share

In research involving people, one may face individuals and groups who think and
act on the basis of values that deviate from the views of right and wrong most people
in our society have. In extreme cases, these may be criminals or terrorists who see
no wrong in killing, or people who, on religious grounds, support actions that are
considered illegal and morally reprehensible. Researchers who, on the basis of their
own moral values, cannot accept or respect such strongly divergent mindsets must
nevertheless show respect for the fundamental rights of these people, as they are
enshrined in the United Nation Declaration of Human Rights and in national law.
They must be treated fairly and correctly alongside everyone else. Their thoughts,
utterances and actions must be studied and discussed without prejudice on the basis
of events and circumstances that have shaped their background. At the same time,
the research must be carried out in such a way that it cannot help to legitimize ethically unacceptable statements, actions and opinions.

This also applies to research on historical figures. In times when society's values
were completely different from today, it is disrespectful to assess the opinions and
actions of the deceased up against today's social norms and values. From a scientific
point of view, this can quickly lead researchers astray.

13.17 Caution when Researching Other People's Motives

In research on what people say, write and do, interpretation of utterances and actions will usually form a central part of the work. In such cases it often becomes important to understand the *motives* behind people's opinions and actions. In a research project this can be one of the most uncertain things to cast light on, partly because the motives are not always expressed, partly because people do not always tell the truth about their own motives. When people's motives are drawn into research, the researcher often enters the borderland between the scientific and the speculative, where many questions regarding methodology and ethics can arise. The starting point in research should be to find facts that are able to indicate to a degree of probability what the motives may be, or that can clarify the truthfulness of what people state as their motives. The uncertainty related to this must, of course, always be discussed thoroughly. However, making assumptions and theories beyond what the facts strictly show also has a place in science, provided that it is based on sound scientific practice and research ethics.

In a research ethics context, the challenge here is twofold. First, it is scientifically challenging to make assumptions and theories about the motives of others in a way that does not violate research ethics truth norms. In general, it is a question of clearly distinguishing between facts and assumptions, clarifying the uncertainty, discussing alternatives, etc. Second, it can be easy to do injustice to people by assigning them motives they do not have. This also applies to the dead, whose legacies may be harmed. People's motives must also always be treated individually, i.e. one must avoid stereotypical perceptions, for example, that bosses' motives are power, companies' motives are money, politicians' motives are to be re-elected, etc. Motives must also be discussed in objective and neutral terms, i.e. without judgemental or glorifying formulations. The underlying circumstances of people's motives also form an important part of the truth about their opinions and actions.

13.18 Caution when Researchers and Their Research Subjects Become Close

In some research projects, researchers can get very 'close' to those who are the subject of their research. This can happen through frequent and prolonged personal contact; when the research concerns personal matters; when it is easy to sympathize with the research subjects; when the researcher also has other roles in relation to the person who is the subject of the research (for example as a doctor), etc. Sometimes, research methods are used where involvement is the whole point, as in so-called action research (see for example Action research (n.d.)).

Closeness can trigger various forms of sympathy or perceptions of friendship, which can both undermine the researcher's objectivity and influence or change the expectations of the person participating in the research. Such relations can develop both mutually and unilaterally. Good practice indicates that researchers are aware that this can happen, and continuously have a critical eye on themselves and those involved in the research in order to capture and possibly avert the development of personal sympathies or relationships between researchers and participants (individuals or groups). The problem here is thus twofold: First, closeness can increase the possibility that research is influenced in an unbalanced, skewed and biased direction, which violates recognized research ethics norms. Second, closeness can trigger unacceptable personal relationships between researchers and research participants.

References

Action research. (n.d.). *Wikipedia*. Retrieved September 21, 2021, from https://en.wikipedia.org/wiki/Action_research

Council for International Organizations of Medical Sciences. (2016). *International ethical guidelines for health-related research involving humans* (4th version). CIOMS. Retrieved September 21, 2021, from https://cioms.ch/wp-content/uploads/2017/01/WEB-CIOMS-EthicalGuidelines.pdf

DLA Piper. (n.d.). *DLA Piper's data protection laws of the world handbook*. Retrieved September 21, 2021, from https//www.dlapiperdataprotection.com/

European Union. (2016). Regulation (EU) 2016/679 of the European Parliament and of the council of 27 April 2016 on the protection of natural persons with regard to the processing of personal data and on the free movement of such data, and repealing directive 95/46/EC (general data protection regulation) (text with EEA relevance). Brussels. European Union. Retrieved September 21, 2021, from http://data.europa.eu/eli/reg/2016/679/oj

Fossheim, H., & Ingierd, H. (2015). *Internet research ethics*. Oslo. Retrieved September 21, 2021, from https://press.nordicopenaccess.no/index.php/noasp/catalog/book/3

Franzke, A. S., Bechmann, A., Zimmer, M., Ess, C., & The Association of Internet Researchers. (2020). Internet Research: Ethical guidelines 3.0. Retrieved September 21, 2021, from https://aoir.org/reports/ethics3.pdf

Office for Human Research Protections. (n.d.). *International compilation of human research standards*. U.S. Department of Health & Human Services. Retrieved September 21, 2021, from https://www.hhs.gov/ohrp/international/compilation-human-research-standards/index.html

The National Committee for Medical and Health Research Ethics. (2009). *Betaling til deltakere i medisinsk eller helsefaglig forskning*. [Payment for research participants in medical and health research. In Norwegian]. Oslo. The Norwegian National Research Ethics Committees. Retrieved September 21, 2021, from https://www.forskningsetikk.no/retningslinjer/med-helse/betaling-til-deltakere-i-medisinsk-eller-helsefaglig-forskning/

The National Committee for Research Ethics on Human Remains. (2019). *Guidelines for research ethics on human remains (English version)*. Oslo. Retrieved September 21, 2021, from https://www.forskningsetikk.no/en/guidelines/social-sciences-humanities-law-and-theology/guidelines-for-research-ethics-on-human-remains/

United Nations (1948). Universal declaration of human rights. Paris. United Nations. Retrieved September 21, 2021 from https://www.un.org/sites/un2.un.org/files/udhr.pdf

World Medical Association. (2013). *WMA declaration of Helsinki – Ethical principles for medical research involving human subjects (2013 version)*. World Medical Association. Retrieved September 21, 2021, from https://www.wma.net/policies-post/wma-declaration-of-helsinki-ethical-principles-for-medical-research-involving-human-subjects/

Chapter 14
Research in and on Other Cultures

In certain research fields and projects, the work focuses on gaining knowledge about foreign cultures or on comparing selected issues in different cultures. This type of research raises specific ethical issues. While it is not possible to cover every aspect of this in this book, some examples may be appropriate:

- Certain types of research may be accepted in the researcher's home country but prohibited or ethically unacceptable in other countries. An example may be the collection of biological material from old cemeteries. Under certain conditions this is acceptable in many countries and cultures (see Sect. 13.15), but not in others. Researchers who do research on human remains abroad must then both follow the relevant ethical guidelines in their own country and relate to the laws and ethics in the country in which they are working.
- In foreign countries and cultures, people who get involved in or are influenced by a research project may react in a different way than is usual in the researcher's home country. Certain types of research can, for example, be more burdensome for participants in some foreign countries. An example could be women's studies in countries where women, for religious or cultural reasons, may find it more stressful and perhaps dangerous to participate as an informant or interviewee than women experience in the researcher's home country.
- For various reasons, certain cultures and subcultures (special religious movements and sects, ideological movements, national or ethnic minorities, etc.) – both abroad and in the researchers' home country – may oppose research or research on certain issues that affect them. This is problematic because in principle it should be possible to study all issues, everywhere and in all societies, with scientific methods in order to uncover facts and develop understanding. Allowing people who have particular opinions about research limit that research violates fundamental principles of academic freedom and the search for truth. At the same time, people's opinions should not be ignored, but taken into account in the design of the research projects, i.e. within the room for manoeuvre set by

D. Slotfeldt-Ellingsen, *Professional Ethics for Research and Development Activities*, https://doi.org/10.1007/978-3-031-25484-0_14

scientific, legal and ethical considerations. The task is to reach compromises that minimize conflict.

- Research on human subjects and the use of personal data usually requires the free, informed consent of the persons involved; see Sects. 13.5, 13.6, 13.7, 13.8, and 13.9. In some countries, cultures and groups, this can present practical and methodological problems for various reasons. The rules for giving consent and obtaining any prior approval of the project (if needed) may also be different than in the home country.

- In countries with widespread corruption, abuse of power, lack of freedom, religious fanaticism and the like, it is conceivable that research and research results may be misused in unexpected ways – a situation which is unpredictable. For example, informants who have participated in a social science research project may be subject to reprisals from those in power who do not want opposing opinions to be expressed. Or that a research collaboration may be misused as propaganda to legitimize a controversial regime. The individual researchers' ethical responsibility for the consequences of their own research requires that the possibilities for such unknown consequences are thoroughly assessed in connection with the planning of research abroad.

'Respect' is the keyword in research ethics when it comes to how research in and on other cultures should be conducted. In reality, people often think, talk and act on the basis of personal and prejudiced notions of cultures that are different from their own – a bad habit that is as old as humanity. One regards others' customs and social behaviour in an unnuanced and condescending way and lets negatively charged words such as primitive, underdeveloped, corrupt, totalitarian, barbaric, etc. characterize their description. However, the opposite is not uncommon either. Some people and researchers tend to be attracted to what is different, and uncritically glorify the customs and societal order of other cultures. Both are doing injustice to other cultures and can be seen as an expression of disrespect. The history of science has several examples of researchers who were more out to support a biased view of a foreign culture than to develop a balanced picture of it. Such research violates several research ethics norms, but the wrong can still be difficult to see because prejudices – which may be unconscious – may be kept hidden and because such biases in research can be obscured or explained away in different ways.

Research in and on cultures that do not share the same views on *fundamental human rights and legality* as in the home country is particularly challenging. Respecting such cultures must mean that one treats them objectively, fairly and without prejudice, but not that one accepts the realities of their view of humanity or opinion on the law.

Because people can develop prejudices gradually and imperceptibly, it can be useful to start new research work on foreign cultures with a self-examining review to bring out the positively and negatively charged opinions or assumptions one may have. Conversations with colleagues who have different views can also be elucidating. Early in the planning process, one should also enter into a dialogue with representatives of the foreign culture to be studied – ordinary people, authorities and local research colleagues alike – to clarify any questions of methodological and ethical nature that should be taken into account in project planning.

What has been said here about research on other cultures will to a large extent also apply to studies of cultures and types of society *back in history* which are different from our own society today.

Within research ethics, as discussed earlier in the book, there is an increasing emphasis on the research being of benefit to society. Society expects to 'get something in return', both for the funds society spends on research and, where relevant, for any personal participation in the research. With this way of thinking, it is natural to plan projects in and on foreign cultures so that those who in some way become participants or are influenced by the research also get something in return. This can, for example, be realized by providing assistance for the follow-up or the use of the research results locally as a final work package in the project. In projects where this is not appropriate, other measures may be considered.

Chapter 15
Research Involving Animals

15.1 Increased Focus on Animal Welfare

Humans have always had to relate to animals, but the view of how animals can be used and treated has developed in step with the development of human culture. Today, animal ethics is about balancing the need to exploit animals in various ways for the benefit of humans against consideration for animal welfare. The modern view is that animals are sentient beings having intrinsic value regardless of the usefulness they may have for humans. They must be treated well and protected from unnecessary pain, suffering, or harm (see for instance the Global Animal Law GAL Association (2018)). Modern societies consider this so important that the main guidelines for the treatment of animals are regulated by national laws. The laws vary from country to country, but generally have provisions that are relevant both for research conducted on animals in experiments and other research involving animals. For animals in experiments, there may be detailed separate regulations.

15.2 The '3Rs'

The view of animal experiments is largely based on three internationally recognized principles, the '3Rs';

- *'Replacement'*: Animal experiments should be avoided – where possible, such experiments should be replaced by alternative research methods. The vision is that experiments on animals will eventually become unnecessary with the help of new experimental methods.
- *'Reduction'*: The number of animals in animal experiments should be limited to the absolute minimum in order to achieve the purpose of the experiment.

© The Author(s), under exclusive license to Springer Nature
Switzerland AG 2023
D. Slotfeldt-Ellingsen, *Professional Ethics for Research and Development
Activities*, https://doi.org/10.1007/978-3-031-25484-0_15

- *'Refinement'*: The methods used during animal experiments should be improved to avoid or minimize that the animals feel pain or fear, or are harmed or subject to other permanent damage. This also applies to breeding, keeping and caring for animals.

The 3Rs was first proposed by Russell and Burch (1959) and became one of the foundations for legislation and ethical guidelines. The regulations are in gradual development as respect for animals as independent individuals and an understanding that animals can feel pain and anxiety grows. New technology also raises new professional and ethical issues. Increased use of genetically modified animals in experiments is an example of this. The main professional and ethical challenge in the years to come, however, is probably not related to adjustments and extensions of the regulations, but in the implementation of them: a real reduction in the use of laboratory animals; stricter assessment that the projects are truly so important, useful and feasible that they justify the use of laboratory animals; better methods to reduce the impact on such animals, etc.

Despite the rigorous regulations, which generally have broad support, there are few things that can arouse such debate and emotions as human's treatment and use of animals. The limit for what is ethically justifiable to expose animals to for research purposes will therefore always be up for debate, both within and outside the research community. Disagreements sometimes lead to unacceptable actions, and from time to time this also goes beyond legal research on animals. Legal producers, sales organizations and users of laboratory animals, for example, have been subjected to direct attacks. Anyone who legally researches animals should therefore reflect on the controversial ethical issues of their activity, partly to be sure that they are able to ethically justify their research and partly to be prepared for criticism or, in the worst case, violence.

References

Global Animal Law GAL Association. (2018). *UN convention on animal health and protection (UNCAHP) (1st pre-draft)*. Zurich. Retrieved September 21, 2021, from https://www.global-animallaw.org/downloads/Folder-UNCAHP.pdf

Russell, W. M. S., & Burch, R. L. (1959). *The principles of humane experimental technique*. Methuen.

Chapter 16
Professional Writing

16.1 Standards of Integrity in Professional Writing

Scientific writing puts researchers in a position where they face some of the most absolute demands for honesty and openness that exist in their profession. It is in the description of their own and others' scientific work that researchers can and must fulfil three of the key requirements of research:

- *Truthfulness.* This means, among other things, that the authors must stick to the facts, present all the facts (including those that contradict their own theories or conclusions), give a balanced presentation, and use formulations that do not mislead the reader.
- *Traceability.* This means, among other things, that the authors must provide the reader with information about ideas, data, arguments, formulations, figures, etc. they have taken from other sources, and what those sources are. Similarly, they must account for how their own data, observations, etc. have emerged.
- *Verifiability*: This means, among other things, that the authors must provide all the information others need to control or repeat the work – as far as possible.

In the same way that scientific authorship constitutes the realization of good research morality, it is also during the writing process that poor quality, sloppiness and dishonesty are made concrete. Here, plagiarism has a distinctive position because in a research context it is almost always linked to the writing process itself and also occurs more often than we like to believe (plagiarism is dealt with in more detail in Sect. 22.4.4).

The three requirements listed above primarily apply to scientific articles, project reports (scientific and administrative), presentations at scientific conferences, etc. but it is also expected that the information in books, popular science articles, etc. written by researchers, is truthful and reasonably traceable.

D. Slotfeldt-Ellingsen, *Professional Ethics for Research and Development Activities*, https://doi.org/10.1007/978-3-031-25484-0_16

16.2   Publication of Research Projects and Results as a Tool in Research Ethics

Research is only useful when the results are made known to other researchers and wider society. New knowledge is usually most widely disseminated when research results are published through articles in scientific journals and at conferences. The results of research leading to commercial inventions achieve the same when the invention is published in the form of a patent. In the context of research ethics, publication is also a key tool for documenting that research has been carried out in accordance with relevant codes of conduct for research. The main reasons why research results should be published are:

- In order for as many people as possible (researchers and others) to benefit from the research results. Among other things, this contributes to increasing the pace of the development of new knowledge and understanding.
- In order for others to subsequently be able to control or verify the research, and possibly detect and correct any scientific errors or breaches of responsible research practice.
- In order not to spend resources on repeating in whole or in part research that others have done before (which one is not aware of because the works have not been published).
- In order for the foundations for research-based decisions that affect people to be made available.
- In order for researchers to be able to document their academic merits and experiences.

The first three bullet points in the list above are mainly arguments for sensible resource utilization and quality control. The last point is often the most important for researchers personally.

At the same time, in many contexts, there are good reasons for why research results cannot or should not be published. For example, significant parts of the total R&D activities in all countries are carried out primarily to meet the needs of organizations or companies. This applies in particular to the business sector's R&D activities, where, as mentioned earlier, it would be unreasonable to give competing companies insight into R&D results that have commercial value (the business sector accounts for a large part of the world's patent literature, but only a few per cent of the total scientific publication). The same also applies to many public bodies where for a variety of reasons certain R&D results cannot be published, for example in defence and security research. Within the university and institute sectors, there may also be good reasons for not publishing a research result or for postponing publication or adapting it to the circumstances. Examples are when the research results contain information about people that one should or is obliged to treat confidentially (discussed in several places in Chap. 13), in contract research (discussed in Sect. 9.4) and in the commercialization of research results (discussed in Chap. 18). In recent years, ethical issues related to researchers' publishing strategies and choice of publishing channels have also emerged (discussed in Sect. 16.9).

16.3   Good Research Practice for the Use of Other Sources in One's Own Written Work

16.3.1   Use of Others' Selection of Topics and Layout of the Text

The content of scientific articles, dissertations and reports is often organized in special ways that have developed over time. A scientific journal article, for example, typically consists of a summary, introduction, description of method, description of the work performed, discussion and conclusion. The structures vary somewhat depending on the format and purpose of the work and from subject to subject. In some cases, organizations, journals, etc. have also established more detailed guidelines for how the material should be organized, which one must then follow. These layouts for the presentation of scientific texts can not be attributed to individuals, and everyone can therefore use them and adapt them to their own needs without referring to the source.

With books, popular science articles, etc. where there is no well-established template for organizing the content, this is a different matter. Here, the organization of the content, the selection and emphasis of material, etc. constitutes a central and often original part of the book's or article's intellectual content. Everyone is free to draw inspiration from or even copy the organization and selection of content from a previous, comparable work, but good practice dictates that this should then be mentioned and the source stated, for example in a preface or separate section containing thanks and acknowledgements. However, many scientific books, especially textbooks, contain approximately the same material and there may be few alternative, logical ways to organize the text. An author who writes such a book without following any specific model may end up with approximately the same content and layout as in a previous book of which the author may not even be aware. This is not plagiarism, but can from time to time lead to unpleasant accusations of being a plagiarist.

16.3.2   Use of Text, Images, Tables, Etc. from Other Sources: Quotation, Paraphrase and Mention of Other Sources

All research is based on previous research. In scientific and professional writing it is therefore very common to use text, figures, data, etc. from previous sources. In order not to violate the norms of truthfulness for scientific writing, clear perceptions have developed over time about how this should be done in practice, more specifically related to rules for quotations, paraphrases and references to material from other sources. These rules are not specific to research, they apply to all authorship, and the first training in good practice starts at school. This is otherwise an area where good practice is primarily learned from others, i.e. from good textbooks and scientific articles.

Despite the fact that the rules for quoting, paraphrasing and mentioning are elementary learning, a number of cases of plagiarism in research have demonstrated that many researchers have insufficient knowledge about these rules and about the boundary between negligence and dishonesty. Remarkably, this also applies to senior managers and heads of research. For this reason, good practice in this area is described in great detail in the following.

Quoting

- A quotation is an exact repetition of a text from another source.
  - Quotations are used to retrieve statements from other sources that are particularly well-formulated and relevant.
- Absolute requirements: When quoting other sources, make it very clear to the reader:
  - That what is written is a quotation from another source.
  - Where the quotation starts and where it ends.
  - Where the quotation is taken from (full reference; see later).

These requirements apply regardless of the content of the quoted text, where it is taken from, or how much is quoted.

The following example satisfies these absolute requirements for marking quotations (a short excerpt of text from an online encyclopaedia that everyone can understand, Ystad (2015), is used here, translated from Norwegian by the author):

Example: In a scientific treatise on the Norwegian playwright Henrik Ibsen, a young researcher wants to substantiate her research through a quotation from a well-known Ibsen researcher. She, therefore, begins her dissertation with:

'Henrik Ibsen is one of the world literature's greatest and most influential playwrights. He brought renewal to the ancient tragedy in the form of realistic plays in prose in which ordinary people take on the role of the classical tragedy heroes, thus adding new artistic vitality and clear societal significance to the bourgeois drama. With his symbolist plays, he also anticipated the development of twentieth-century expressionist and modernist theatre art.' This is how Ibsen's significance is summed up by Vigdis Ystad, one of Norway's foremost Ibsen researchers (Ystad, 2015)

In this example, the quotation (marked with quotation marks) is placed directly in front of the text to create flow. At other times, and especially with long quotations, it may be more appropriate to separate the quotation from the text by making it a separate paragraph. One can also use methods other than quotation marks (for example, italics) to mark what is being quoted. This is a matter of *style and typography*. What is important is that the quoted text differs in a clear way from the rest of the text and that the reference is provided together with the quotation

In a short text consisting of generalities, there may be a significant probability that a self-written text may coincide with something that another has written before:

Example: The literature researcher in the example above could also express herself as follows:

Henrik Ibsen is one of world literature's greatest and most influential playwrights. His plays have been compared to the classical Greek tragedies, and the purpose of this work is to analyse the similarities and differences in content and structure between Ibsen's dramas and the classical Greek tragedies.

The first line here is exactly as in Ystad's article but is also a generality that any literary scholar could have written without having read Ystad's text. When something one writes completely or partially coincides with something others have written (of which one does not know), it naturally cannot be marked as a quotation or paraphrase. However, if the literary scholar had written the text above after reading and been inspired by Ystad's text, she should in principle have marked the first line as a quotation. However, since this is a short sentence of generalities that everyone could happen to write, it is quite common not to mark the quotation. The prerequisite must then be that it is a short piece of text that is trivial when it comes to content and wording. However, this is a grey area where people hold different opinions, at least if the text is somewhat longer than in this example

The main rule when quoting is that the text must be reproduced word for word and letter for letter. This applies to everything – language form, spelling, grammar and punctuation. However, when a quotation is integrated into a new text, it may be appropriate to adjust a little to create flow, facilitate readability, clarify the meaning, correct typographical errors, etc. It is therefore acceptable to make *small* adjustments (but as little as possible) and only on the condition that these are carefully marked. For example, if one chooses to omit some words in a sentence being quoted, these omissions should be marked with [...] (possibly only..., which is particularly common to use when omitting something at the beginning or end of the sentence). And if one is adding words, these should be marked with [own words]. The same applies if, for example, an underline, italics etc. in the text are removed or added. Different style guides have slightly different ways of marking this.

Examples: The literature researcher in the example above could find it necessary or appropriate to adjust the quotation in different ways and mark it as follows:

Additions:

Henrik Ibsen '... brought renewal to the ancient tragedy in the form of realistic plays in prose in which ordinary people [from our time] take on the role of the classic tragedy heroes, thus adding new artistic vitality and clear societal significance to the bourgeois drama' (Ystad, 2015)

Omissions:

Henrik Ibsen '... brought renewal to the ancient tragedy in the form of realistic plays in prose ... and thus added new artistic vitality and clear societal significance to the bourgeois drama' (Ystad, 2015)

Typographic changes:

Henrik Ibsen '... brought renewal to the ancient tragedy in the form of realistic plays in prose in which ordinary people take on the role of the classic tragedy heroes, thus adding new artistic vitality and clear societal significance to the *bourgeois drama* [my emphasis]' (Ystad, 2015)

The requirement that the quoted text must be reproduced verbatim also applies in principle to misspellings, typos, etc. in the original text. Based on common sense, however, this may be unnecessary:

Examples: Suppose that 'Ibsen' in Ystad's original article had been misspelt 'Isen'. A true purist who quotes from this text will then keep the error in the quote, but inform the reader that there is an error in the original text. It can be done as follows:

'Henrik Isen [sic] is one of world literature's ...'

Alternatively, the error can be corrected but marked clearly:

'Henrik I[b]sen is one of world literature's...'

However, unless the typo can be said to have a bearing on the perception and character of the quoted text, it is difficult to see anything significantly wrong in correcting this type of error *without* any marking:

'Henrik Ibsen is one of world literature's ..'

For all quotations:

- *Both* quotation marks and reference must be used.
- The reference must always be in *direct* connection with the quoted text. There should be no doubt as to where the quoted text is taken from, or what the reference concerns. Thus, it is, as an example, wrong to refer to an earlier source in the introduction of an article and then quote from this source later in the work without marking the quotation and repeating the reference.
- The reference to the source should refer to the exact place from which the quotation is taken. This means, for example, that the page number must be given when referring to a book. Correct ways of referring are described in Sect. 16.4.
- In principle one should refer to original sources and not quote a text that is in itself a quotation. If the secondary source is used to find the primary source, it is good practice to refer to both. The reference that follows the quotation can then, for example, be formulated as follows: 'From NN (reference to the secondary source) who quotes MM (reference to the primary source)'.
- All quotations and references in the document must be marked in the same way. Changing style is not acceptable because it can create confusion about what are quotations.
- Quotation marks are not only used to mark quotations but also to mark a word in a foreign language, an unusual way of wording, certain names, oral speech, etc. Based on the context of the text and the possible absence of a reference, such use of quotation marks can hardly be confused with a quotation.

Quotation marks and references must be used regardless of what kind of content the quoted text has, or where it is taken from. Mistakes are made from time to time. For example, some researchers have the misunderstanding that citation marking and reference to the source is unnecessary when:

- The quotation is taken from documents which:
  - Are not 'scientific' or not 'research'.
  - Can be used 'freely' (such as text from encyclopaedias, websites, public documents, etc.).
  - Are their own works.

- The quoted text:
  - Concerns well-known material in the research field.
  - Is banal (below the 'threshold of originality'), or not covered by copyright law (more on this in Chap. 17).

Any text that is so well worded and has such relevant content that one wishes to quote it verbatim should therefore never be presented as one's own formulations, but be marked as a quotation in accordance with good practice. Good practice for citations thus protect the quoted *formulations* themselves and the original *selection* of content in them.

Quotations from one's own works must also be marked. The main purpose is to clarify the chronology of the texts and enable others to trace the origin of everything in the work. If there are also different co-authors on the documents, transparency considerations about who has contributed what will also be an important part of the justification.

The use of quotations has become more and more common because articles etc. are now available online in digital form, and because modern word processing tools make it easy to 'cut' or 'copy' text written by others and 'paste' it into one's own manuscript. At the same time, it has become easy to make changes to copied text. However, new technology does not change the standards for professional writing. An action does not become any less wrong simply because it has become easier to carry out. The diligence due to marking quotations and referencing sources is therefore just as necessary now as before.

Paraphrasing

- A paraphrase (also termed an indirect quotation) is a reworking, reproduction, summary or similar of the content and meaning of a text, formulated in one's own words. In a research context, the paraphrase shall be a truthful reformulation of the original text.

  - Paraphrasing is usually used to give a brief restatement of the content or essentials of another source.
  - The advantage of a paraphrase over a quotation is that it can be more easily adapted to the text's style and content flow, and that it can be made short with a focus on what is important in the original text.

- Absolute requirements: When paraphrasing texts from other sources, make it very clear to the reader:

  - That what is written is a paraphrasing of another text.
  - Where the paraphrasing begins and where it ends.
  - The source of the paraphrase (full reference; see below).

This requirement applies regardless of the content of the paraphrased text, where it is taken from, or how extensive the paraphrase is.

A paraphrase is usually formulated in the same written style as the rest of the text and in such a way that it naturally blends in with the flow of the text. In order for the paraphrase not to be confused with a quotation, quotation marks or other ways of distinguishing the paraphrase *graphically* from the rest of the text are not used. In order to satisfy the requirement that the reader should not be in doubt about where the paraphrase begins and ends, one must then usually be quite explicit in the text that there is a paraphrase. The simplest and most common is to start with a text indicating that the part that follows is a paraphrase of another text, and to end the paraphrase with a reference, as illustrated below:

Example: If the literary scholar in the examples above wanted to use of Ystad's original text, but shorten it all, she could make a paraphrase in which the main points were summarized:

V. Ystad describes Henrik Ibsen as one of world literature's most influential playwrights, an innovator of classical tragedy, who revitalized contemporary art and anticipated the development of modernism in the art of theatre (Ystad, 2015).

The text at the beginning of the sentence marks sufficiently clearly the start of the paraphrase, and the reference at the end marks its ending. The text contains the same main points as the original, but is shorter. That it is a paraphrase and not a quotation is shown by not using quotation marks. This procedure can also be used if the paraphrase contains several sentences between the introduction and the reference.

For long paraphrases, such as an extensive summary of a passage in other source, one can mark the beginning and end of the paraphrase by placing it as a separate paragraph or chapter in the text, and through the introduction to the paragraph or the wording of the headline make it clear to the reader that it is a paraphrase.

The most difficult part of paraphrasing is to reproduce the original text in a truthful way, i.e. neither omit important content nor add more than the original contains.

Example: If the literary scholar in the example above is not careful to choose words that cover the meaning of the original text, she ends up with a false paraphrase:

V. Ystad describes Henrik Ibsen as one of world literature's most influential playwrights, an innovator of classical tragedy, who revitalized contemporary art and created modernist theatre (Ystad, 2015).

This paraphrase satisfies the absolute requirements of a paraphrase (the reader understands that it is a paraphrase, and sees where it starts and ends), but the text is not entirely true to the original. Through an unfortunate choice of words in the last part of the paraphrase, the reader gets a slightly wrong impression of Ystad's view of Ibsen's significance for modernist theatre (there is a difference between anticipating something new and creating something new). In this example, the unfortunate wording is obviously the result of sloppiness and poor craftsmanship. Done intentionally and to a much more serious extent, however, untrue paraphrasing can be viewed as falsification or fabrication. It is seriously wrong, for example, if someone else's text is deliberately presented so that the reader can believe that the person being paraphrased supports the author's own views to a greater extent than is the case.

Paraphrasing requires an understanding of the nuances of the language and careful control of the words used to make sure that they express the essence and true meaning of the original text. Having said that, it is important to be aware that a paraphrase will never express *exactly the same* as the original and that it will be a subjective element in the assessment of what is a true restatement. Everyone who rewrites must therefore be open to criticism – a risk that a conscientious author may take.

It must be emphasized that a paraphrase should be a real rewriting of the original. Making cosmetic changes and using large parts of the original text verbatim, is quoting, not paraphrasing. Presenting a quotation as a paraphrase is a violation of good practice and can be seen as plagiarism.

Mentioning

- A mention is a description of or reference to specific contents of previous works or other sources. The purpose is to:

  - Describe elements of one's own work in a short way.
  - Specify results, observations, data, etc. that are obtained from other sources and used in one's own work, and give recognition to the originators.

- Absolute requirement: When mentioning other sources, make it very clear to the reader:

  - Which source one is referring to (full reference; see below). This requirement applies regardless of the type and content of the source.

In research reports and journal articles, the work of others is typically referred to and discussed in:

- The introduction and the literature review, where it is possible to significantly shorten the description by referring to others.
- The description of methods, where it is possible to significantly shorten the description by referring to others who have developed or previously used and described the methods.
- The description of the research work where results, data, etc. from other sources are often used.
- The discussion of the results, where results, data, etc. from other sources are often compared to one's own results.

16.4   References

Reference to the source is absolutely central to all quotations, paraphrases and mentions. The reference should be unambiguous and make it easy for the reader to find the place in the original source used. A reference must state the name of the authors, the title of the work, the name of the journal, publisher or similar, any booklet number, the year, page number and anything else needed for the source to be unambiguously defined. References and reference lists can be designed in many stylistic ways, but the style chosen must be used consistently. Most journals and publishers require that the references must be written according to specific style templates, such as Chicago style, Harvard style, APA style (American Psychological Association), etc. The choice of style often follows the tradition in the research field and the research environment. Research organizations may also have preferred style templates that students and staff should or must follow. The most common style templates are readily available online. All of them are equally good when it comes to satisfying guidelines for research ethics related to making references

16.5   Writing Style, Spelling and Grammar

Good scientific work is characterized by clear and important issues, well-considered methods, accurate studies, logical and neutral argumentation and sound conclusions. This concerns the *content* of the work. When the work is to be *described* in the form of a scientific article, dissertation, scientific report, popular article or similar, it is important to use a style (language, typography, etc.) and spelling and grammar that reflects the quality one strives for in the research itself. Good practice for scientific writings is then to use a *unified style* when it comes to language and typography and *consistent and correct spelling and grammar* throughout the document. A scientific document that appears in a clear and correct external form makes it probable to the reader that the author places high demands on quality. The opposite – a changing, shabby style, and sloppy and unclear language – can be an indication that the author may have been similarly inaccurate and sloppy in the research work itself. It is also not difficult to find dissertations and research reports with examples of such poor disposition and language that the reader can easily be misled about the content and conclusions of the work. The emergence of new language forms on mobile phones and in social media on the internet, the growth of multicultural societies, widening gaps between oral and written language, use of dialects, etc. also affects people's general ability to express themselves in a way that everyone is able to comprehend on an equal footing. Research organizations' quality assurance should therefore increasingly also emphasize linguistic clarity in the description of research.

All researchers *can* make themselves write correctly and clearly. Not doing so, therefore, is a matter of negligence. In the most serious cases, this could be seen as a breach of the code of conduct for research and an expression of bad research morality – a researcher should never be sloppy.

16.6   Criteria for Co-Authorship

16.6.1   Scientific Works (the Vancouver Recommendations)

Today, research is a *team effort* to a large extent. No one in the team contributes to an equal extent, neither in the actual research work nor in the writing in connection with the reporting and publishing of the work. The question then arises, who qualifies to be listed as authors of reports and articles? This issue is also relevant in the relationship between student and supervisor.

Through authorship, one assumes responsibility for the work that has been done and may take the praise for the results. Research merits are mainly achieved through scientific publications. The number of publications (co-)authored, where the works have been published and the attention the publications have received (measured by the number of citations or in another way) are all emphasized. This is important not only for the individual researcher's career but also for the organization where they work (the number of publications is the basis for measuring research productivity in many countries and research organizations, and public financial support may to some degree be based on this). At least at the beginning of their professional careers, most researchers experience great pressure to publish – the publications are decisive for their careers ('publish or perish'). In such a context there will also always be someone who tries to cheat to obtain a higher score.

In order to prevent cheating and promote fairness, it is therefore important to have clear criteria for what it takes to become a co-author. The criteria have traditionally varied somewhat between different fields of research and may differ from country to country. In recent years, however, it seems that the research communities *in all fields of research* have coalesced around the Vancouver recommendations (see Sect. 2.4.3); see box below. This is important since much research today is international and interdisciplinary.

The Vancouver Recommendations (i.e. the ICMJE Recommendations) for Co-Authorship
In order to be a co-author of a scientific publication, the following four criteria must be met (International Committee of Medical Journal Editors, updated December, 2019):

1. Substantial contributions to the conception or design of the work; or the acquisition, analysis, or interpretation of data for the work; AND
2. Drafting the work or revising it critically for important intellectual content; AND

(continued)

3. Final approval of the version to be published; AND
4. Agreement to be accountable for all aspects of the work in ensuring that
   questions related to the accuracy or integrity of any part of the work are
   appropriately investigated and resolved.

In addition to being accountable for the parts of the work he or she has
done, an author should be able to identify which co-authors are responsible
for specific other parts of the work. In addition, authors should have confidence in the integrity of the contributions of their co-authors.

All those designated as authors should meet all four criteria for authorship,
and all who meet the four criteria should be identified as authors. Those
who do not meet all four criteria should be acknowledged ...

The criteria are not intended for use as a means to disqualify colleagues
from authorship who otherwise meet authorship criteria by denying them
the opportunity to meet criterion #s 2 or 3. Therefore, all individuals who
meet the first criterion should have the opportunity to participate in the
review, drafting, and final approval of the manuscript. (from 2. Who Is an
Author)

Any national versions of this may be slightly different.

It is important to note that, for example, contribution to funding, general management and supervision, technical and administrative support, editing and proof-
reading, etc. do not *alone* qualify for co-authorship.

In addition to these recommendations, the practice of many publishers and others
is that the authors must explain in writing their contributions to the research work
and the publication.

Those who become co-authors according to these recommendations can list the
work on their CV as documentation of their own scientific work. The Vancouver
recommendations apply primarily to scientific journal articles. R&D reports that are
not published as journal articles will generally also be of professional merit to the
researchers, and the Vancouver recommendations on co-authorship should therefore
also be followed for such reports.

Based on criterion four of the Vancouver recommendations, some believe
that all the co-authors are equally to blame if one of them commits scientific
misconduct. However, being responsible for the work as a whole does not mean
that one is automatically guilty of what a partner does wrong. As described later
in Sect. 23.1, everyone must be assessed individually and it is for example generally far more reprehensible to commit scientific misconduct than not to discover that a partner has cheated. In order to have some protection against unfair
accusations, the collaborators might consider specifying in detail the specific
responsibilities of each project participant, preferably in the collaboration
agreement.

16.6.2 Common Forms of Violation of the Co-Authorship Guidelines in Scientific Publications

Three forms of unjustified co-authorship have been quite common (Nylenna, 2015):

1. 'Gift authorship' is a situation in which persons with a relatively tenuous association to the project are included on the list of authors, perhaps in the hope that they will reciprocate the favour next time around.
2. 'Guest authorship' is the term used when particularly well-known or prominent persons are unfairly invited to be included on the list of authors because it is assumed that this will strengthen the project and increase the chance of publication.
3. 'Ghost authorship' is the term used when persons who definitely should be included among the authors are omitted – willingly or unwillingly. (Excerpt from the section named The Problems)

The first two, often referred to as 'honorary authorship', can be seen as a form of cronyism, or worse, of giving others credit for a scientific work that they do not deserve. Maybe in order to thank the person for support of some kind, maybe to achieve benefits in the form of collegial friendship, rewards, increased prestige, attention, etc. There have traditionally been many cases of this in science, technology and medicine, and the Vancouver recommendations are an important tool to stop this practice, which still prevails to a certain extent. In the humanities, social sciences, law, and other field, the problem has traditionally been the opposite. Here examples of elements of 'ghost authorship' are probably more common.

In recent years, various forms of violation of the rules for co-authorship have also been linked to new forms of scientific misconduct and forgery: Some seek to 'launder' articles that cannot be defended scientifically or that are pure fake science (see Sect. 22.4.6), by enticing well-known researchers to co-author the articles (a form of gift authorship). To accept 'co-authorship' to unscientific or false articles should objectively be described as an act of scientific misconduct, while the degree of guilt must be assessed on a case-by-case basis.

16.6.3 Special Issues for PhD Dissertations and Master's Theses

A PhD dissertation is usually either a monograph or a collection of independent publications (or preprints) beginning with a chapter that binds it all together. The latter goes under several names: article dissertation, cumulative dissertation, compilation dissertation, etc. A master's thesis is usually a monograph.

In the case of monographs, the PhD student is the sole author of the dissertation, even though today's doctoral students have carried out their work under supervision, and the supervisor may have contributed a good deal. The supervisor's (and others') contributions are then described and honoured in the preface or under a separate section with thanks and acknowledgments. The same applies to master's theses.

In the case of article-based dissertations, the introductory or summarizing chapter must always be written by the PhD student alone, while the independent publications may have a supervisor and others as co-authors. Only the PhD student's name is listed on the front page of the dissertation. With regard to independent publications included in an article-based dissertation, there are field-dependent differences in the view of how the supervisor's contribution should be assessed:

- In some fields, such as science, technology and medicine, the supervisor is almost always co-author of these publications. Here, it is common to regard the supervisor's contribution, for example, to the idea and design as an intellectual contribution to the work that should form the basis for co-authorship. The prerequisite, however, is that the contribution is significant, and that all four criteria in the Vancouver recommendations are met.
- In the humanities, social sciences and more, the opposite is often the case: the supervisor is most often not included as a co-author. Here, it is common to regard the supervisor's contribution as guidance, while concrete contributions, for example to the idea and design of the work, are mentioned in a preface or separate section for thanks and acknowledgments.

These differences are due to different traditions within different subject areas and countries.

Many master's theses do not provide enough original results to warrant being published in a scientific journal. It is then quite common for the *supervisor* to compile material from several master's theses and other research projects, which are then eventually published. The supervisor then becomes the prime author, while the master's students and others who have made a sufficient contribution are given the opportunity to be co-authors. Unfortunately, from time to time supervisors choose not to include their former students in the publication process, even though the students have contributed sufficiently to qualify as co-authors. The fact that the publication is prepared after the students have left university or college may present practical difficulties but does not exempt the supervisor from including former students. Unjustified omission as a co-author is, however, not only poor conduct towards a student but a clear deviation from good research practice. Such cases occasionally appear at research organizations, and in these cases the threshold for concluding plagiarism and/or other breaches of the codes of conduct in research is relatively low.

16.6.4 Special Practice for Patents

Writing a patent is a highly specialized task that is most often performed by a patent office or a patent expert, based on the inventors' documentation of the invention. The finished patent is published in the owner's name (often a research institution or company, more exceptionally the inventors themselves), while the inventors are named as inventors. Who authored the patent is not stated and is not relevant. However, a patent is highly meriting for the inventors, and in those cases where the patent is based on teamwork, the question arises as to who qualifies to be named as inventors. Since patents may also generate revenue, it is important that this issue is handled properly. Companies and research institutions where patenting occurs regularly should therefore have their own guidelines here. The use of neutral advisors, preferably external, in the assessment should also be considered.

As with co-authorship of scientific articles and reports, it is reasonable that only those who make *significant contributions* to the invention are given the status of inventors.[1] The contributions to the very basis of the idea and the original solutions – the specifically *patentable* – will then be decisive, while contributions in the form of work to test, demonstrate and document the invention should not in themselves suffice for inventor status. Furthermore, contributions to financing, general management and supervision, technical and administrative assistance, etc. will not give inventor status. The latter may seem obvious, but within commissioned research, it happens that the client mixes ownership rights and inventor status. From time to time this can lead to conflict (un example was given in Sect. 9.5.3).

16.6.5   Special Practice for Less Meritorious Notes and Technical Reports, Administrative Reports and More

Scientific publications and research reports are not the only documents from the hands of researchers for which authorship criteria are an issue:

- *Unpublished working notes, task reports, etc. in R&D projects.* This refers to documents that are usually prepared during the project work, for use internally in the project group, or as information for management, clients and others. If the material here is of a predominantly scientific nature, it is natural that such documents are written by the researchers involved. It is then appropriate that the co-authorship is decided in accordance with the Vancouver recommendations, whenever possible. One reason for this is that such notes and task reports can be important in documenting who are the originators of scientific discoveries or technological inventions.
- *Project plans and applications, status reports and annual reports/final reports for R&D projects.* These documents are usually prepared in the name of the organization formally responsible for the project. They often both contain

---

[1] Various aspects of an inventor's rights are protected through national legislation relating to patents, employee's invention rights and copyrights.

some scientific information (descriptions of the project idea, plan for or summary of the scientific work, etc.) and information of an administrative, organizational, financial and bibliographic nature. Sometimes the information is split up into separate reports. Such documents will not normally be of scientific merit to the researchers involved, and the Vancouver recommendations are then less relevant. However, they can be managerially and administratively meritorious for those who take the initiative, plan, lead and/or have overall responsibility for the project. It is natural that the project manager writes many of these documents on behalf of the organization, regardless of how much the manager has participated in the research work (the research organization may have its own routines for this). In order to avoid misunderstandings about who is responsible for what, it is then also common to give an explicit account of the project's organization, management, staffing and division of tasks in such documents. Any initiators and others who have not participated in the actual project work should also be indicated, including any clients or others from outside who have contributed intellectually to the project.

• *Reports in connection with assignments that are not R&D (consulting work, engineering, technical assistance and analysis, surveys, etc.).* The persons who perform the work are often crucial in assignments like this. Examples are consulting and advisory assignments where the client buys services from researchers on the basis of their expert knowledge. The clients usually hold them professionally responsible for the work and the results, even when the contract has been entered into with the research organization. It is then natural that the reports are written by those who have carried out the assignment. Where several have participated to varying degrees, an adapted version of the Vancouver recommendations may be suitable as criteria for co-authorship.

In other commissioned projects, the research organization is in focus for the client. The names of the employees who perform the work then have no meaning, as long as the work is performed professionally. This is typical of commissioned laboratory assignments of various kinds. Here, the project manager appears as the natural report author. Sometimes other solutions may also be appropriate, and the research organization may have its own procedures for this (for example, that the author's name is not given, only the name of a contact person or the manager responsible for the assignment). When developing such procedures, the research organization should consider establishing a practice that makes reports of this type meriting for the key employees in the task – not necessarily scientifically – but managerially and professionally.

In all these cases, the author's name or role in the project and in relation to the document should in one way or another be stated or alternatively accounted for in a special section on organization, management and division of tasks.

16.6.6   Authorship Order

When there are several co-authors of an article or report, the question arises in what order they should be listed. No generally accepted standard exists. The practice varies according to traditions within different fields and research environments. Essentially, there are two main ways to arrange the order of authors:

- The order reflects the role and efforts of the co-authors. Two variants are common:

  - Primary focus on the *role* the co-authors have had in the work:

    - The first author then becomes the person who has had the main responsibility for the practical work, who has performed a significant part of the work and/or who has written the work, and to whom inquiries regarding the work should usually be directed.
    - The last author is often the most senior researcher in the collaboration, who may be the initiator of the work and may have acted as a mentor or supervisor for other co-authors. Senior researchers often contribute in particular to choosing methods and procedures but are usually not responsible for the main effort in the research (but enough to qualify for co-authorship in accordance with the criteria discussed above).
    - The other co-authors are placed between the first and last author, the most important first, but often somewhat random.

  - Primary focus on the *extent and importance* of the co-authors' work:

    - The order is determined on the basis of an objective quantification (as far as possible) of each co-author's contribution to the research reported. Different types of contributions (to the idea and project plan, the literature review, the methods, the data collection, the discussion of the results, the writing, etc.) can be given different weight. This is often done by each co-author first evaluating both their own and the others' contributions based on criteria agreed upon in advance. On this basis, the co-authors then jointly agree on a listing order (simple computer programmes to calculate this can be downloaded from the internet). Increasingly, journals and others require co-authors to make fairly detailed statements about their contributions to the work. This can also form the basis for discussions about authorship order.

- The order says nothing about the role and efforts of the co-authors. In this case, it is most natural to list the co-authors in alphabetical order.

In practice, the first author gets much more attention than the other co-authors. The first of these two alternatives contribute to a logical correspondence between the one who receives the most attention and the one who has contributed the most to the work.

Because practices are so different, co-authors should explain the principles that underpin the authorship order. This can easily be done in a footnote or similar. It is also good practice to give a more explicit account of each individual co-author's contribution to the research work and the article.

16.6.7 Simple Measures to Avoid Disagreements and Conflicts About Co-Authorship

From time to time, disagreements may arise between researchers about co-authorship. To avoid this, everyone who plans to collaborate with others (including student and supervisor) should agree in advance on how to report and publish the project results, what qualifies for co-authorship, and in what order authors should be listed. The agreement must be included in the cooperation agreement. Such a plan will of course have to be changed along the way if something unforeseen happens (for example if one of the participants leaves the collaboration, or the work tasks change).

16.7 Acknowledgment – The Custom of Thanking for Contributions, Help and Support

In project reports and publications, it is good practice to insert a short section where thanks and acknowledgements are given to those who have contributed *significantly* to the project. First and foremost, one should mention those who have made important intellectual contributions, such as a colleague, supervisor or leader who may have suggested or initiated the project or given important advice or input along the way (but otherwise not participated enough to become a co-author). Furthermore, assistants and others who have carried out a major work should be recognized, such as a laboratory engineer who has analysed samples. Finally, those who facilitate the project by making resources, materials and equipment available should also be recognized, including everyone who has contributed to the financing of the project.

The question of who should be given such recognition is a matter of discretion. The criterion for the assessment should be that the contributions should be *significant and important* for the project. Some people make a mistake here by taking the opportunity to name-drop or flatter – they thank the famous Harvard professor for an 'inspirational conversation' (which lasted twenty minutes) and the head of the department for 'invaluable support' (she is nice to everyone). Others go the opposite way and thank the whole spectrum of secretaries, assistants, helpers and students for 'tireless work', generously and collectively regardless of differences in contributions. All this is wrong because it distorts the realities – the requirement of truthfulness and accountability also applies to the acknowledgement section.

16.8   Requirement to Account for Any Actual, Potential or Perceived Conflicts of Interest

Guidelines for research ethics require that researchers unsolicited provide information on matters that may be relevant in the assessment of their neutrality and impartiality in the work. The information must be given to partners, sponsors, funders, any customers, etc., and be included in a relevant form in project proposals, reports, scientific articles, conference lectures, peer statements, media posts, etc. The way the information is provided, and the selection, scope and detailing of it, must be adapted from case to case, but the responsibility for ensuring that sufficient relevant information about any possible conflict of interest is available lies with the researcher.

It will, then, almost always be relevant to provide the following:

- Information about who has financed or otherwise supported the project work and whether it has been carried out on commission or request.
- Information about the project participants and the research organization that may have special interests related to the research work itself or to persons and organizations involved in or related to the project, for example, information about:
  - Project participants who may have close, personal relationships with other persons or organizations that are important for the assessment of their impartiality and neutrality in relation to the project (see Sect. 19.4).
  - Project participants who may derive direct or indirect personal financial benefit from the project.
  - Project participants who may have personal interests, jobs, positions of trust, etc. in political, ideological, religious or other organizations that are relevant to the research.
  - Any long-term collaboration between the research organization/researchers and clients or stakeholders that may be relevant for assessing the researchers' neutrality and impartiality in the research project.

It may be appropriate here to remind readers that there is basically nothing wrong with having personal interests in something if one is open about it (see Chap. 7), and that people's trust in researchers most likely grows when they are open about their interests.

16.9   Ethical Issues Related to New Forms of Publication

Publications are highly meritorious for all researchers. This is an underlying driving force when researchers publish their work in scientific journals or at conferences. The most prestigious journals and conferences count the most. In recent years, new internet-based publishing tools such as online access and open access, the emergence of a large number of new online journals (from high quality to fraudulent), changes in the various journals' funding and publishing strategies, and national systems for measuring research productivity have resulted in major changes in the international scientific publishing industry as well as in the publishing strategies of individual researchers, research organizations and national research authorities. The changes have both scientific and research ethical implications. Five issues in particular have been the subject of discussion:

16.9.1   Transition to Online Journals – A Paradigm Shift in Scientific Publishing

The internet has made it possible to publish research results in a completely different way than in traditional paper-based journals. This has opened up new opportunities for making money from publishing scientific articles based with new business concepts adapted to the internet. In a short time, this has led to the emergence of thousands of commercial online journals in all disciplines. The Directory of Open Access Journals, an independent database managed by a community-interest company based in the United Kingdom, lists close to 17,000 journals (September 2021) that offer free, open access and satisfy certain quality standards (Directory of Open Access Journals, n.d.). The new business idea is most often based on five components:

- There is *online access* to the articles – the internet is the most accessible and efficient platform for the distribution of information. The cost of publishing and distribution is minimal.
- There is *Open Access* to the articles – the authors usually pay the journals to have their articles published ('Article Processing Charge, APC'), while readers do not pay anything.
- The editorial processing time is *very short* – the articles that are accepted are usually published on the internet a few weeks after they have been received by the editors.
- The articles are quality assured by the editorial staff and through a peer-review process before they are accepted for publication.
- There are essentially no restrictions on the number of articles that can be published – the internet has no shortage of space, but there will be practical restrictions, and quality-conscious online journals will reject articles that do not meet the journal's quality standards.

In principle this offering is very attractive to researchers and research organizations as well as to public bodies responsible for research. In the light of research ethics, however, some aspects of this new business concept are problematic.

Since there is no shortage of space on the internet, and when publishers charge the authors to publish an article, publishers make more money the more articles they receive and accept (some charge for evaluating the article). Herein lies an inherent temptation also to publish articles of low quality and originality, and with little content of any importance.

Many researchers also believe that the short processing time these publishers offer – which is a crucial aspect of the business idea – makes a sufficiently thorough editorial and peer-based quality assurance impossible. They suspect that the peer review these publishers advertise is, at best, superficial. For publishers, this can provide increased income and profitability. The fact that these issues are real is beyond any doubt:

- Many researchers have observed or experienced that the threshold for accepting an article in some of the new, commercial, online journals is very low in terms of scientific content and originality. Short studies, experiments and calculations, scientific trivialities, more data on matters that are all well known, etc., are published as journal articles, well wrapped in plausible text. Some will see this as an unfortunate development – the scientific literature can be flooded with work of insignificance. Others – who have particularly criticized the traditional, subscription-based publishers for their strict and skewed selection of article proposals – will applaud the innovation – all research results should be published, important or otherwise.
- Poor standards of quality assurance have been revealed in many of the new online journals (examples are given in Sect. 19.2.1). If many online publishers are sloppy with their peer reviews, the system as a whole will be corrupted and 'peer-reviewed' can no longer be used as a synonym for quality. Many researchers also receive personal offers from online publishers that are of such a nature that it lends suspicion to the idea that something is wrong.

In addition to the many commercial online journals, there are also a number of non-commercial ones that neither charge the authors nor the readers (funding is arranged differently). These do not have a direct commercial motive for publishing many articles, but for competitive and financial reasons may choose to go lightly on the editorial and peer-based quality checks. Therefore, new online journals, commercial and non-commercial, should be treated with some caution.

Any co-author of a scientific article is responsible for the article being published in a journal with well-documented standards of quality that clearly satisfy the requirements most researchers set for editorial and peer-based quality assurance. However, in the myriad of online journals, it can be difficult to distinguish trustworthy from untrustworthy publications and publishers (see later). However, in countries where research authorities and research organizations use publications to evaluate the productivity and quality of researchers and research organizations, there are lists of legitimate journals. Researchers who choose to publish in these can hardly be criticized, even though there are different views in the research community about the quality of some of these journals. Internationally, there are also online lists of both legitimate and predatory journals.

To meet the competition from the new online journals, many of the traditional, paper-based publishers have also developed internet-based solutions. Some then offer open access a certain time after the articles have been published, some take a minor fee from readers for each article they download from the internet, some offer an annual subscription. Research authorities in many countries hope and trust that the most reputable paper-based journals will end up offering open online access.

16.9.2   Open Access to Publicly Funded Scientific Articles

Before the internet existed, paper-based publishers had been given a central role in the research process by being held responsible for deciding which articles to publish under their own quality assurance and selection criteria. The research community became increasingly concerned about the problems with this, among other things because it was possible for publishers' article selection procedures to lead to 'publication bias' (discussed later). The long processing time of many traditional publishers, and the business idea that the publishers' revenue is obtained from the readers through journal subscriptions, was also criticized. The latter is due to the fact that researchers in less resourceful institutions or parts of the world have limited access to scientific literature for reasons of cost. Many consider this unacceptable for ideological/political reasons. In addition, paper-based solutions appear to be impractical compared to electronic ones.

This, together with the rise of the internet, has led to a new, government-influenced, international publishing policy with the objective of giving everyone free, open access to *publicly funded scientific works*. The aim is to be realized in two main ways:

- *'Gold open access'*: By urging or requiring researchers to publish the results of publicly funded research in journals that give *readers* free access to the articles. There are two main alternatives here:

  - To publish in journals that acquire revenue from article authors (the authors pay the publishers to process and publish the articles in the publishers' online journal), while readers get free access to the articles.
  - To publish in journals that are funded by the public sector, by institutions or otherwise, where neither the authors of the articles nor the readers pay anything.

Variants of these also exist.

- *'Green open access'*: By urging or requiring that researchers, research organizations and publishers deposit a copy of published articles, dissertations, etc. in a national or institutional online and searchable repository or archive to which everyone has free access. Different forms of self-archiving on internet sites are variants of this.

In Europe, the implementation of this policy has been facilitated by cOAlition S, an initiative launched in 2018 by a group of national research-funding organizations supported by the European Commission and others. The goal of their plan (named

Plan S) was that all scholarly publications with a certain type of public funding should be available to everyone free of charge (gold or green open access) from 2021 onwards (cOAlition S, n.d.). Although many support the principle of free, open access, there are different views on how it should be achieved. Plan S has therefore received criticism and many countries and organizations in and outside Europe have not immediately supported the plan. One of the objections from the research communities has been that the transition must take place in a way that makes it possible to publish in the most prestigious journals, even if these do not have free, open access. It has also been pointed out that the choice of journal must be part of researchers' academic freedom. However, support for and implementation of the plan is increasing.

The individual researchers lives up to the ideal of openness in research when they describe their research in an accountable, complete and truthful way in scientific articles, project reports, contributions at conferences, etc. In relation to the *norm of openness* in research ethics, publishing in the old way in traditional, subscription-based paper journals is, therefore, an 'equally moral' course of action as publishing with free, open access in internet-based journals. However, choosing a publishing channel with free, open access can make the research results more widely available, more quickly, not least to researchers and research communities with poor financial resources (but at the same time it becomes more expensive for them to publish their own works). Open access can therefore be seen as an ideological and political research tool, which naturally also has its basis in ethical values.

16.9.3 Responsibility to Avoid Fraudulent Publishers and Conference Organizers

As mentioned above, a forest of so-called predatory online journals has grown up in parallel to the emergence of high-quality online journals; see Predatory publishing (n.d.). They deceive readers and writers by pretending to be something they are not, or through false or misleading information about how they operate. Most of them try to appear as legitimate journals but cheat in different ways to generate greater income. Many try to deceive authors and readers by giving their journals credible names (often almost identical to famous journals) and use editorial boards with well-known names to cover up their scam. Some scientists are obviously so vain that they are susceptible to this form of 'honour' without exercising cautious about what they are getting involved in. Most predatory journals advertise editorial and peer-based quality assurance. In reality, they will often publish almost anything, and the peer review is either non-existent or superficial. The purpose is to save money on editorial work and make money by publishing as many articles as possible without their true intentions being revealed.

Purely fraudulent journals also exist ('fake journals', 'pseudo journals'): Some are non-existent 'journals' that entice researchers to pay for publication that never takes place. Some offer to create articles for researchers for a fee, some publish 'research' articles that are not based on scientific methods, etc.

Predatory journals are not uncommon. They are found in the thousands, and tens of thousands of researchers all over the world submit articles and publish in them. Because the articles are not (adequately) quality-assured, no one can trust their quality. On the surface, there is little to distinguish between predatory and legitimate journals. The number of predatory journals is therefore a matter of debate.

Simultaneously with the rise of predatory journals, a new business has developed based on the arrangement of fake conferences ('predatory conferences'): Well-known researchers are duped to participate and are offered honourable roles (opening lectures, chairmanship of sessions and panels, etc.) – with the fraudsters playing on the researchers' vanity. Some are uncritical and easily deceived – they sign up, pay the participation fee and travel to the designated location, only to discover that there is no conference, or that the other participants come from completely different disciplines or the like.

Fraudulent journals and conferences have become so commonplace that it has aroused interest internationally among investigative journalists; see the box below.

Investigative Journalists Shed Light on Predatory Journals and Conferences

In 2018, an international group of investigative journalists from renowned media organizations conducted a survey of five major internet publishers identified as publishers of pseudo-scientific journals: WASET, OMICS, Science Publications, ScienceDomaine and IOSR Journals. Among other things, the journalists examined 175,000 articles written by researchers around the world and published by the five publishers.

To prove that the publishers were engaged in fake science, some of the journalists submitted fictitious 'scientific' articles. One of these was written by journalists at *Aftenposten*, a major Norwegian newspaper (Lundgaard & Strøm, 2018). They used SCIgen to create the article (SCIgen is a computer programme developed at MIT that generates 'scientific' nonsense articles in computer science, which can be used to detect fraudulent journals). The journalists listed their own names as authors of the article but stated a fictitious Norwegian university as their affiliation. They submitted the article both to a 'scientific conference' in computer science organized in Vienna by the online publisher World Academy of Science, Engineering and Technology (WASET) and to one of the approximately 700 'scientific' journals of the online publisher OMICS:

• The WASET conference in Vienna advertised reliable editorial assessment and peer-review. Nevertheless, the nonsense article was *accepted* for presentation at the conference, and the journalists were even sent certificates for participation in advance. After paying the participation fee of €500,

(continued)

they showed up in Vienna and found there about fifty thoroughly deceived researchers who had submitted articles to a number of conferences on different topics, from Renaissance literature to 4D printing for the production of biomaterials. Everything was held in the same room at the same time, led by the same man. WASET's business idea was thus to announce thousands of conferences which were then held simultaneously in one place as a 'multidisciplinary conference' – a meaningless concept that WASET could make money on because, among the thousands of researchers in the world, there are always some who allow themselves to be tempted and deceived. According to the journalists, WASET planned 157 multidisciplinary conferences in 2019, each of which would include 1112 thematic conferences.

- The OMICS journal also advertised reliable editorial assessment and peer-review. Nevertheless, the nonsense article was *accepted* for publication despite the publisher pointing out that over 50% of the article was *plagiarism* that needed to be corrected. The publisher requested US$ 2019 to publish the article in the journal (the journalists stopped the experiment there).

The journalists from other countries who participated in the investigation had similar reports.

In a leader in Aftenposten magazine, the editor, Lillian Vambheim, commented that '... the scope [of fake research] is widespread ... and can have dangerous consequences for our trust in research'. It is of little help that this is condemned by research leaders and well-known researchers around the world. The research community must implement concrete measures to end all publication in dubious journals and conferences.

Researchers who agree to sit on editorial committees and participate in peer reviews in predatory journals or conferences, who are co-authors of articles sent to such journals or conferences, who refer to articles from them, etc. – are in practice contributing to cheating. This requires that the research organizations investigate the cases and react with measures and sanctions against those involved. The phenomenon of predatory journals and conferences is now so well known that only in special cases should researchers claims of being deceived be mitigating.

Everyone has an independent responsibility to investigate whether the journals they use or work for, and the conferences they attend, are of sufficient quality, i.e. that the journal and the conference are genuine, and that the editorial and peer-based quality assurance is satisfactory and according to good research practice. A simple online search for information about the publisher, the journal or the conference, contact with members of editorial committees and a *critical look* at what one finds will often be enough to expose the nonsense.

16.9.4   Disputed Procedures for Increasing the Number of Publications – 'Least Publishable Unit'

Researchers have always competed with each other, primarily to be the first to make great discoveries. In recent times, there has also been a competition to have the greatest number of publications, and research authorities, research organizations and academic publishers have made arrangements to 'organize' and intensify this competition. The number of publications has become a key criterion for ranking and rewarding researchers and research organizations (some would say that it has become the main criterion). High scores on productivity have, for example, become important when applying for positions, scholarships and research funding. Many researchers have therefore begun to adapt their publishing strategy accordingly.

> Example: A researcher wants to study the differences in ways and means for integrating immigrants in the five Nordic countries. The project is set up as a collaborative project between him and a researcher in each of the other countries. The team agrees to divide the project into six work packages. First, they want to collect facts related to the situation in each country and publish them as one article for each country (each with the national researcher as the main author and the four others as co-authors). Then the team will make a comparison between countries – which is the real aim of the work – and publish this in a separate article (with the project manager as the first author and the four other national researchers as co-authors). In this case, however, the first five articles will contain little new because the ways and means of integrating immigrants are well known from other studies and evaluations. However, the work needs to be done in order to have a unified set of data that can make a comparison between the countries possible. The last article, on the other hand, will contain new and interesting observations. The scheme was chosen to get six publications out of a project that 'in the old days' would probably have generated a single publication (everything could have been published in one, slightly longer article).

The researchers divided the work into sub-projects and speculated in what has been called the *'least publishable unit' (LPU)*, i.e. the scientifically thinnest article that can be published in a journal or at a conference with peer review (Least publishable unit, n.d.). This publishing strategy can be problematic for both scientific and ethical reasons, but especially because it makes it easy to publish insignificant research. The predatory journals and many of the new, legitimate, online journals accept 'thin' articles, i.e. they operate with a very small LPU.

In addition to researchers deliberately writing LPU-based articles in order to get many publications, some also seek to become *co-authors* of as many publications as possible. Small contributions on many projects can 'pay off' rather than a lot of work on one project because a long list of articles makes a CV more impressive than a short one. In many contexts, cooperation, and in particular cooperation with other countries, is also often prioritized by those who fund research. Such incentives may tempt researchers to seek collaboration with others where with a minimum of intellectual contribution and work effort they can become a co-author of an article. This is possible because the criteria for co-authorship contain a discretionary assessment that the contribution must be 'substantial' (see Sect. 16.6.1) – this can be interpreted strictly or liberally. In parallel with the term LPU, one can therefore talk about a lower limit for how much one needs to contribute to become a co-author – a kind of 'least acceptable author contribution (LAAC)'.

LPU and LAAC do not really represent anything new when it comes to concepts. What is new is that the boundaries of what is acceptable enough have changed and thereby opened up new opportunities for researchers to merit themselves in terms of the number of publications. Today, it is not uncommon for researchers to be co-authors of a dozen articles or more per year (two years in a row, a professor at a Norwegian university co-authored well over 100 articles per year in legitimate journals (The National Commission for the Investigation of Research Misconduct, 2017)). This is hardly because researchers are more skilled and industrious than before. In many disciplines, new technology and new forms of collaboration within research can obviously explain the increased output, but the increase is probably also a result of the fact that in certain respects the overall scientific literature is undergoing a change of character.

In practice, LPU and LAAC are defined by the research community itself (researchers suggest which journals they want to be considered acceptable, the list of acceptable journals is managed by representatives of the research community, and the research community defines the criteria for co-authorship). From a research ethical standpoint, LPU and LAAC are by definition limits to what the research community finds *academically and ethically acceptable*. Researchers who publish 'thin' articles in journals with a low publishing threshold can therefore not be criticized on moral grounds, at least not if the journal is accepted as qualifying for credits by the research community. The condition must be that the possibilities of having scientifically 'thin' articles accepted, do not lead researchers to choose research tasks that are scientifically or socially less important or useful. Using national research resources and public research funds in this way should be considered irresponsible. Many therefore believe that ploys to get as many publications as possible out of a research project are unacceptable in terms of research ethics. Others believe, on the other hand, that adaptation to changes within certain limits cannot be wrong. Striving for many publications is, after all, a result of institutional and national research strategies and has the backing of many actors, from institute leaders to cabinet ministers.

16.9.5   Unwanted Publication Biases

Surveys have revealed that in recent years many studies in medical and biomedical research have not made it to publication. Analyses of the multiple causes indicate that studies where positive effects are found, or which confirm previous studies, are prioritized by the publishers. Studies that show no effect, or that contradict previous studies, are on the other hand downgraded. The result is that the overall research literature gives a skewed picture of the research that has actually been carried out. The phenomenon is called 'publication bias'. This can happen in any field of research. The reasons can be many: Experience has shown that positive findings attract more attention than negative ones, and publishers can therefore consciously or subconsciously prioritize the studies that give the journals the most attention. Studies that unsettle established knowledge can also easily be viewed with more scepticism and be subject to stricter assessment by peers and editors than more mainstream studies. Many fear that publication bias delays the development of knowledge and in the worst case can have harmful consequences on a par with skewed reporting of the results in an individual project. Against this, one can argue that there can be nothing wrong with publishers taking quality and commercial considerations into account when choosing which articles they want to print. Publication bias is an issue that primarily relates to the traditional and reputable paper-based journals. In the new, rapidly growing forest of open-access online journals, it is seldom difficult to find acceptable journals that are willing to publish articles that the more prestigious journals have rejected. The problem is rather how to deal with the vast amounts of scientific results that are reported, and how to distinguish important and reliable research from insignificant and untrustworthy work.

References

cOAlition, S. (n.d.). *Plan S Principles*. Retrieved September 21, 2021, from https://www.coalition-s.org/plan_s_principles/

Directory of open access journals. (n.d.). *The directory of open access journals*. Retrieved September 21, 2021, from https://doaj.org/

International Committee of Medical Journal Editors. (2019, December). Recommendations for the conduct, reporting, editing, and publication of scholarly work in medical journals, section II.A.2 who is an author? Retrieved September 21, 2021, from http://www.icmje.org/icmje-recommendations.pdf

Least publishable unit. (n.d.). *Wikipedia*. Retrieved September 21, 2001 from https://en.wikipedia.org/wiki/Least_publishable_unit

Lundgaard, H. & Strøm T. J. (2018, August 17). Forskningsbløffen. [The research bluff. In Norwegian]. *A-magasinet (Aftenposten), 33*, 15.

Nylenna, M. (2015). Authorship and co-authorship in medical and health research (English ed.). Oslo. The Norwegian National Research Ethics Committees. Retrieved September 21, 2021, from https://www.forskningsetikk.no/en/resources/the-research-ethics-library/authorship-and-co-authorship/authorship-and-co-authorship-in-medical-and-health-research/

Predatory publishing. (n.d.). *Wikipedia*. Retrieved September 21, 2021, from https://en.wikipedia.org/wiki/Predatory_publishing

The National Commission for the Investigation of Research Misconduct. (2017). *Uttalelse i sak om mulig vitenskapelig uredelighet ved universitetet i Agder* [Statement in case of possible research misconduct at the university of Agder. In Norwegian]. Oslo. The Norwegian National Research Ethics Committees. Retrieved September 21, 2021, from https://khrono.no/files/2017/11/15/uttalelse-i-sak-om-publiseringspraksis_anonymisert.pdf

Ystad, V. (2015). Henrik Ibsen. In *Norsk biografisk leksikon* [in Norwegian]. Oslo. Store norske leksikon. Retrieved September 21, 2021, https://nbl.snl.no/Henrik_Ibsen

Chapter 17
Authors' Legal Rights

17.1 General

Authors of scientific works, like others who create original, literary or artistic works, possess a number of rights that are protected by national laws. The laws differ from country to country, but common legal traditions, international conventions, agreements and the harmonization of laws in some countries (for example in Europe) have contributed to much being the same.

There are two components to authors' rights: Copyrights (which are intellectual property rights, i.e. economic rights) and moral rights. These statutory rights are occasionally brought into the argumentation in cases that mainly concern research ethical issues. A brief overview of what they concern may therefore be appropriate in this book on ethics.

17.2 Authors' Copyrights

Copyright is essentially an exclusive right to make copies of an originally created literary, scientific or artistic work, and to make the work available to the public. It is the creator of the work who in the first instance acquires this right, usually at the moment the work is created. The copyright can be waived, transferred, sold, leased, etc. to others. As an example, works created by employees as part of their work duties, on commission for a client, and the like, usually become the employer's or client's property depending on their employment contract. Copyrights are time-limited, with this limit differing from country to country.

The copyright shall contribute to the author being able to make money from the work. To use a copyrighted work or material from it, one must therefore have the owner's permission and possibly pay for the use. But not all works are copyrighted

D. Slotfeldt-Ellingsen, *Professional Ethics for Research and Development Activities*, https://doi.org/10.1007/978-3-031-25484-0_17

(so-called 'public domain works', which have no copyright protection due to expired copyright, little originality, etc.), and copyright laws also allow some limited use under special circumstances without the owner's permission. Four aspects of copyright laws are particularly relevant in connection with scientific writing:

- Scientific articles or reports, dissertations, conference contributions, illustrations, computer software, etc. are 'works' in the meaning of copyright laws.
- Only the concrete *expression of the work* is protected by the copyright. For instance, in the case of a scientific article or report, this would be the selection and disposition of material in the document, the wordings, the design of figures and tables, etc. The scientific ideas, methods, theories, data, etc. are not protected. The reason for this is that the results of research should be available to everyone and that no scientific work should block other works in the same field or with the same methods.
- The work must exceed a *threshold of originality* in order to qualify for copyright. In a text, this has first and foremost to do with the degree of originality in the wording. Parts of a work may meet the standards of originality while others do not. However, the threshold of originality is generally not high, and for precautionary reasons, one should assume that the text in scientific articles, reports, theses, etc. is so special that it qualifies for copyright.
- Anyone can quote texts from a published, copyrighted source provided it is done in accordance with good practice for quoting and to the extent that the purpose requires (but quoting the entire text or major parts of it would be an infringement).

What is described here are just some of the key aspects of copyright law. There are many details and some provisions differ greatly from country to country. Readers who come across copyright issues must therefore look up their national laws.

The last bullet point above is particularly important in research because researchers to a large extent use text, tables, figures, etc. from other sources when describing or reporting their own research. As addressed earlier (see Sect. 16.3), this is acceptable when done correctly. Violation of the guidelines for using text etc. from other sources is, however, in the most serious cases considered research misconduct, i.e. plagiarism (see Sect. 22.4.4). If the sources are copyrighted, this may in addition be a violation of copyright law. However, the criteria for the former are not the same as for the latter. Copyright law can therefore not be used as a basis for determining whether someone has plagiarized in a research ethics context, i.e. violated norms related to traceability, truthfulness, respect for the work of others, and more.

17.3 Author's Moral Rights

The author's moral rights have two basic components:

- A right to be attributed as the author of the work when it is copied and published.
- A right to preserve the integrity of the work, i.e. a right to object to alterations and use of the work that violates the author's honour and reputation.

Moral rights to a work are the creator's personal rights regardless of who owns the copyright to the work. Again, the details of these provisions vary from country to country and the readers must consult their national laws. In many countries, these moral rights are included in the copyright laws.

Chapter 18
Commercialization of R&D Results from Research Institutions

18.1   Commercialization – When the Concrete Societal Benefit of a Research Result Is Realized Via the Business Community

History has shown that research serves society. However, the path from a single research result to concrete application and direct benefit to society can be extremely winding and tortuous. In practice, most research projects contribute small steps towards building up the general level of knowledge (in a broad sense) in various areas. In reality, many research projects have little or no importance for the development of knowledge, while very few form the basis for large leaps. The level of knowledge is also built up in other ways than through research. It is first and foremost through the sum of contributions to the development of knowledge that the research as a whole gradually benefits society. To a considerable extent, this happens through the education of professionals who, at universities, colleges and other places, learn 'the state of the art' in the various subject areas and make use of it when they enter working life.

Sometimes, however, specific research results can also be *used more directly by society*. There are two main mechanisms for this. Within the social sciences, the humanities, law, certain fields in the natural sciences, etc., knowledge and data from one project or a group of projects can often *be used right away* by government agencies, business companies and others. An example could be the results of a research project on juvenile delinquency that can be used directly by the authorities as part of the factual basis in work to establish new and better, preventive measures. In many areas of medicine and mathematics, science and technology, however, the research results only really benefit society when they are used to develop *new or improved technologies, processes, products and services*. That is often a long and far more elaborate process. An example may be the results of a research project in inorganic chemistry which indicates that a hitherto untested lithium-metal oxide

D. Slotfeldt-Ellingsen, *Professional Ethics for Research and Development Activities*, https://doi.org/10.1007/978-3-031-25484-0_18

with a special composition has properties that possibly make it suitable as a cathode material in an electrochemical cell for use in batteries. The knowledge about this is in itself of little use to society. The benefit is only realized when, based on the research results, someone has developed, started production and marketed a finished battery that is competitive with alternative batteries in terms of price and function. Research institutions do not have the expertise, capital and other resources needed to do this themselves. The business community does. Some research results, therefore, do not really benefit society until they are somehow made available to and *used by the business community* – they must be '*commercialized*'. The fact that research results contribute to new business activities is in itself beneficial to society thanks to the revenues and jobs it provides.

18.2   The Business Community's Operating Conditions and Working Methods – An Important Premise for Commercialization

In order for a research result to be commercialized in an effective way, researchers should understand the business sector's situation and way of working – and vice versa. Some factors that are typical of the business community's way of operating are therefore addressed below. There is a particular focus on start-ups and technology companies, where the business is often based on *unique technology*, and companies, where there are *significant investments* behind the products. Some research ethics issues related to this are then discussed.

First, it is important that researchers do not overestimate the importance of their own research results and underestimate the scope and importance of the industrial development that is necessary for the results to be commercialized. A *specific* research result that a company either picks up for free from the open, scientific literature or acquires the right to use in another way is often just one of several starting points or prerequisites the company has for developing something new and saleable, one of many elements in a long development chain where the industrial aspect can be far more capital and resource-intensive than the research behind the results acquired from outside. The *innovative* elements in the industrial and market development are also often significant and crucial. In the example of the cathode material that the chemists discovered above, an industrial company that somehow 'picks up' the result must go on to verify the researchers' findings, optimize the material composition, develop or adapt the other components of the electrochemical cell, develop a cost-effective manufacturing method and find solutions related to the use of the new material. They must also develop innovative, practical solutions for the battery packs they end up offering in the market. Sometimes the researchers are commissioned to participate in this development, other times not. Many times, companies end up with products and processes mainly based on knowledge and technology quite far from the research results they obtained from outside. But sometimes, of course, it also happens that the original research results remain the crucial innovative element in the new industrial development.

Secondly, it is important to understand that most businesses have limited competence and resources to follow the open scientific literature, let alone pick up concrete, new, enabling knowledge or technology directly from it. Today, new research-based technology and knowledge are largely available to ordinary companies *indirectly*, through the purchase of technology, materials, production equipment, parts, etc. from specialized, external suppliers who have the resources and expertise to monitor and utilize the scientific and technological development in their fields of expertise. Some of these companies want to operate close to the research front in their core areas, and often seek direct collaboration with relevant research groups. In turn, researchers working in fields where the results may have a commercial application often want to have direct contact with potential users at an early stage. In such collaborations, the researchers' role is usually to be the company's eyes and ears in the research community. They can scout for new, enabling technology and come up with, test and make trial use of new solutions that may be of interest to the companies. The role of the companies is to turn the innovation into useful products and services from which they can make money and for which there is a good market. They can also give researchers insight into where there is a particular need for more knowledge and new innovations. For this to yield results, both parties must respect each other's roles, possibilities and working methods and enter into sensible and responsible compromises where necessary. However, the 'closeness' between the parties also raises ethical issues that are discussed elsewhere in the book.

Thirdly, it is important for researchers to understand the realities regarding the economy and competition in the business sector. A company starting out on the development of a new or improved technology, process, product or service often has to *invest significantly* (from millions to billions) in the development work itself, in the modification or acquisition of production equipment, sometimes in the construction of production lines or entire factories, in any government-mandated approvals, in the introduction to market, and more. The investment must be financed by the company's savings or borrowed/obtained from public and private sources. All of this is also associated with risk and those who invest in the development will therefore always demand compensation for the risk they take. The greater the risk, the greater the profit potential must be. This is the reality regardless of whether the money comes from public or private sources, and regardless of who owns the company (public, private or mixed ownership). In many cases, the biggest risk is that other companies simultaneously develop equally efficient technologies and processes, or competing products and services. All companies will therefore seek to reduce this risk by ensuring, as far as is necessary and possible, that their own processes and products are based on knowledge and technology that others are not aware of or have access to. Few will, for example, take the risk of spending tens of millions of euros on the development, approval and launch of a new, primarily research-based drug if competitors also have access to the same research results and thus may be able to quickly copy and market the same drug. This is the basis for patenting or keeping certain research results secret.

18.3   Patenting and Secrecy of Business-Critical Information

Patenting and secrecy are the most common tools industrial companies have to reduce the risk that competing companies may launch copy products that may contribute to a decline in the return and profitability of their own products.

Secrecy is used here and elsewhere in the book as a collective term for what can be called 'active' secrecy and different degrees of 'passive' secrecy. In the case of active secrecy, no information is provided about what one wants to keep to oneself; all documents are strictly secured; access to production and development premises is severely restricted; knowledge of formulas and procedures occurs on a need-to-know basis; rigorous measures are implemented to prevent industrial espionage, etc. This is seldom needed. In the case of passive secrecy – which is the norm – one finds it sufficient not to actively go out with information about the company's own solutions and makes do with normal security arrangements that are found in most organizations. In practice, this gives an edge over competitors when developing and marketing something new. But apart from some in-house solutions and know-how, one must expect that everyone in the industry over time will gain knowledge of each other's technology, processes and products. This is because the majority of companies participate in the open exchange of knowledge and technology in society; competing companies also collaborate in many matters; employees change workplaces, etc.

Patenting is only possible if the research results include an *invention*; see the box below. Resourceful research institutions often file for patents before contacting companies. It provides better control over the commercial use of the knowledge, strengthens the negotiating position vis-à-vis companies, and enables greater profits by selling or leasing out patent rights. Patents are published and are therefore also meriting for the researchers who are listed as inventors (see Sect. 16.6.4). However, patenting is expensive, and the risk that the revenue and profit will fall short of expectations is often significant. Many research results may have commercial value for a company without being patentable or patented, but the value then often depends on how many competitors have access to the same.

Patents on Research-Based Inventions – a form of Deferred Publication That May Increase the Likelihood that Research Results Will Be Used Industrially

Research results are rarely in themselves directly patentable. However, if the results contain or can be further developed into an *invention*, i.e. a new and original, practical solution to a problem, then the invention may be patentable. Patents can be obtained for a *specific technical, industrial (in a broad sense) method, product or use.* It is assumed that the invention differs significantly from previous solutions (that the invention has an 'inventive step' or 'non-obviousness') and can be utilized industrially.

(continued)

A patent gives the patent holder control over the use of the invention. Among other things, no one may, without the consent of the patentee, manufacture, sell or use a product that is protected by the patent or use a method that is protected by the patent. In some countries, however, patented inventions can be used freely in certain types of research. The national patent laws differ and readers who either have to use patented inventions in their own research, or believe that their own inventions can be patented, should consult their national patent authorities.

The owners of patents can use them to protect their own manufacturing. They can also give or sell licenses to others. The license agreement will state what rights the licensee receives, and any terms and limitations in the use of the invention. Patents can also be sold to others.

Patents are granted by national patent offices upon application by the owner of the invention. A granted patent usually gives the patent holder the exclusive right to the invention for up to 20 years after the application was filed. Patents must be applied for in each country or in groups of countries. It costs a good deal to both get and maintain a patent (especially if one needs patent protection in many countries). Many, therefore, choose to keep inventions secret.

In order to obtain a patent, the invention must not be made public before the patent application is submitted. Because changes may be required in connection with the application process, it is also common to postpone any publication until everything has been clarified. The patent office will in any case publish granted patents. A patent is therefore a form of postponed scientific publication. In return, the invention is described in great detail.

Laying out a good strategy for how to protect an invention with the help of patents, and how to formulate patents, requires a high level of expertise. Some engage a patent agency or internationally operating consultants who specialize in patenting in various types of industrial and other activities.

A patented invention or a secret research result will generally be more valuable to the industry than a research result that anyone can read about in the scientific literature. However, some still question whether patenting and secrecy are really important prerequisites for industrial companies to invest and take risks. They argue that openly available research results will lead to greater, better and more efficient societal and industrial use of research. In the real world, some research results are likely to be commercialized more easily and quickly if patented or kept secret while others may be more easily and quickly commercialized when they are open to everyone for free.

The need for patenting and secrecy is generally assessed on a case-by-case basis in companies. The decision-makers usually have a good understanding of what they should patent (patenting is expensive, patents can be contested, etc.), what they should keep secret (secrecy is demanding and secrets can be revealed), and what they may be open about. Industrial companies also have far better expertise and factual bases than researchers do to assess the usefulness or necessity of patenting and secrecy in each case. In those cases where researchers are dependent on the business community for the research results to benefit society, there are therefore good reasons to listen to their assessments and choices if one wants the results to be used quickly.

The need to protect oneself by patenting and secrecy varies greatly depending on the individual company's products and services, the scope of investments behind the business, the competitive situation the company is in, etc. In many companies, much of what is said here about patenting and secrecy will therefore be less relevant.

18.4   The Authorities' Expectation that Researchers Should Contribute to Commercialization

The commercialization of research results has taken place as long as research has been conducted, and many of the giants in the history of science are also recognized for their direct contributions to innovations in the field of industry and elsewhere. An example is Lord Kelvin (William Thomson, 1824–1907). While a professor at the University of Glasgow, he became famous for his studies in thermodynamics, as well as for his many inventions. He is credited with 661 scientific articles and 70 patents (William Thomson, first Baron Kelvin, n.d.), and the combination of scientific and industrial merits contributes significantly to making him a great scientist. In recent years, research collaboration with the business community has been institutionalized at many universities and colleges. Today, modern academic institutions have three main missions: education, research and collaboration/interaction with society. A number of technical-industrial research institutes have also been established around the world with industrial competitiveness and innovations as their main mission. The expectation that research will contribute to the development of industry and other fields is embedded in the authorities' research and business policies. Significant portions of public funding for research have industrial innovations – new or improved processes, products and services – and the creation of new jobs as their main goal. Public funding also often supports the commercialization of academic research. Cooperation between business enterprises and research institutions, for example, is central to many of EU's R&D programs. This means that the commercialization of research results has in may fields become a key part of the responsibilities of many researchers and research institutions.[1]

---

[1] It is common for research institutions to acquire the ownership rights to research results and inventions of commercial value created by their employees. The conditions for this are always described in more detail in employment contracts and internal procedures. The responsibility for commercializing the results of research, when relevant and possible, therefore often lies with the research organization. Resourceful research institutions with industry-oriented research often have

18.5   Alternative Procedures for Commercialization

Research-based inventions made at universities, colleges or research institutes (which are not the result of commissioned research for clients) can mainly be commercialized in two ways:

1. The researchers or their research organization can offer the results to one or a selection of companies that they believe have an interest in and the resources to exploit them commercially. The researchers or their research organization then *take an active step to commercialize* the results. This means that the research results are initially kept secret from most people, at least for a certain period of time. Based on what is said in Sects. 18.2 and 18.3, this will often increase the chances that companies take the risk of investing in the innovation. When companies have to pay for the results, they also have an incentive to use them. In the contract with the buyers, the research organization can often also be given a degree of certainty to ensure that the research results are actually used, and control how they are used. For example, they can ensure that licences to use the results do not cover ethically questionable applications. They may also include provisions in the contract that stimulate ethically sound usage of the results. By offering the research results to national companies, they can especially contribute to strengthening and restructure the industry sector in their own country.

2. Researchers can publish the results in scientific journals or in other ways. In principle, this makes the research available to industrial companies all over the world, free of charge and at the same time, and it is thus up to the companies to assess whether and how they can utilize the results commercially. The researchers and their organizations are passive in this. However, they may also contribute somewhat more directly to commercialization by making selected companies aware of the commercial opportunities in the published works and perhaps offering expert help in connection with any commercial development and use.

These are the main choices. There are also variants and combinations of them, but in the following, only the two are considered, with the main emphasis on alternative 1, which stands for what is generally associated with the commercialization of research. Each country and institution will have its own procedures for choosing between these and for the collaboration between the researchers and the institution in these matters.

---

employees, departments or subsidiaries with special expertise and responsibility for commercialization. However, the researchers who create the results will always represent the first step in the commercialization process, with the main responsibility for informing the employer of the commercial possibilities. The researchers, with their professional expertise, are also often a valuable resource in the work of transforming research results into useful industrial products and processes.

18.6   Some Research Ethics Issues in Commercialization

Based on society's expectations of researchers (see Sect. 6.3), it is natural to look at the commercialization of research results primarily as a tool for realizing a moral obligation to society, i.e. as a tool for realizing concrete benefits of certain types of research. With such a point of view, the choice between alternatives 1 and 2 then becomes a question of choosing the procedure that best contributes to this happening in practice. Some of the most important factors to consider in this regard are discussed above. However, the two alternatives include different approaches that raise some obvious research ethics issues. Four of them will be discussed here:

The first one has to do with the researchers' individual responsibility for the consequences of their own research, which was discussed in Sect. 6.5. In connection with active partaking in the commercialization of research results (alternative 1 above), this responsibility will then first and foremost include:

- Being cautious in selecting partners for commercialization. In the same way as in commissioned research, collaboration with untrustworthy companies or companies that are involved in illegal or ethically unacceptable activities must be avoided.
- Assessing whether the products or services that may result from the collaboration with the company can be defended ethically. This will then become a follow up of the impact assessment made before the research project was initiated (see Sect. 6.5).

The second has to do with disclosure versus secrecy. Alternative 1 generally means that parts of the research results are not published immediately and that something may be kept secret for a longer period of time. Some believe that this is unfortunate in light of a number of research ethics norms. In addition, many emphasize that all research is based on previous research throughout the world and that researchers therefore have a moral obligation to make their own research results available to others in the research community. However, researchers also have a moral and legal obligation to those who fund the specific research in question (often public funding bodies), and society has interests and needs that must also be safeguarded.

Many public R&D programs have business development as a goal or sub-goal. The contract terms for supporting research projects then often balance between meeting the business community's needs to patent or keeping certain research results secret, and the research community's interests in publishing. In some cases, thus, the authorities find it more appropriate for the research community to contribute actively and purposefully to developing concrete commercial opportunities than to develop the open, common knowledge base. The fact that the scope of research results which, due to commercialization, cannot be published (or cannot be published immediately), constitutes a very small part of the scientific production, also has a certain significance here.

The third concerns who should have ownership of research results. Alternative 1 means that the research results are initially reserved for some selected companies.

As discussed above, the purpose is to help reduce the financial and technological risk companies must take when the research results are to be used as a starting point for developing new or modified processes or products. However, some believe – usually on a more principled basis and based on their own ethos and opinions about the nature of research – that research results should be freely available to the public, and should not be a commercial commodity. However, the prevailing view in society is different. In all countries, the results of a research project are regarded as an intellectual property (IP) that can be owned by researchers, research institutions, companies and others. An owner of an IP has a large degree of control over it and can use it or sell it, protect it through patent or secrecy, etc. In practice, large portions of the research results still become public property because publication is the main principle, and only a small proportion is commercialized directly through individual companies.

The fourth has to do with the possibility that potential income from the utilization of IP can unreasonably tempt researchers and research organizations to make wrong priorities when it comes to choosing research topics and using research resources. This issue is relevant because most researchers will probably think that the choice of the research projects should be determined by what is scientifically important or socially useful, not by what provides the greatest financial return. The ethical concern here, however, is not that research results are a commercial commodity, or that researchers and research institutions prioritize research areas where the results may contribute to innovations that society can benefit from, but that the *revenue potential* is used as a criterion (there is no general proportionality between important and useful research and the revenue potential of a research result). In practice, the concern is probably exaggerated, most researchers are predominantly guided by the interesting, important and useful aspects of their work.

In addition to this, a number of issues that are primarily not related to research ethics often come up in discussions about the commercialization of research results. The discussion about patenting of life that arose a few years ago is an example of this. Herein also lies a more principled question of how to draw the line between discoveries and inventions. However, these are primarily questions related to patent law and patent ethics. Another example is arguments that research results can be abused or only become beneficial for selected groups or interests in society because the business community operates commercially. The business sector is so diverse and extensive that there is no shortage of examples of this. The question here, however, is first and foremost of a political and business ethics nature that society seeks to deal with in many ways through national and international measures, regulations and control. Such and similar issues that occasionally arise when commercialization of research results are discussed, but which primarily cannot be elucidated on the basis of research ethics, are not discussed in further detail here.

With all commercialization, it is also a question of making money. This in itself raises ethical issues that apply more generally than just in research. An example: Many people are concerned with fair trade, i.e. that the price the buyer pays should be perceived as 'right' in relation to what the buyer receives. The assessment of what is the right price can vary considerably, often based on people having different values.

> Example: A researcher at a research institute finds a new way to improve an existing, commercial product. The idea comes by chance, and it is a simple matter for the research institute to test that it works. The product in question became known to the researcher through a previous project for a company, but there is otherwise no connection between the previous project and the new idea. The institute now addresses the company and offers the idea for improvement at a price that is many times greater than the *costs* the institute has had in developing and testing the idea. The researcher reacts to this. He finds it unethical to demand significantly more than the costs have been. This is a not uncommon way of looking at things among researchers, who may think more idealistically than commercially. The management of the institute, for its part, believes that the price must be based on the *value* of the knowledge and technology they sell and base their negotiations with the company on that. It all ends with the company buying the idea at a price that is in good agreement with the institute's assessment, and which thereby gives the institute a very good profit in relation to the costs it has had. Both parties are satisfied, and the research institute gets fresh funds to spend as it pleases. Most people will probably also take a positive view that research institutes in this way make money on research-based innovations.

18.7 New Business Concepts in the Business Sector Can Open New Avenues for the Commercialization of Research Results

There are great differences between companies in the business sector. For many companies, the conditions and prerequisites for collaboration with the research communities on the commercialization of research results can be very different from what is described above. The internet and other digital communication and information channels have, for example, made possible new forms of commercial activity, where new ways and means for commercializing research results may be necessary. In the future, this will probably raise new ethical issues. Ideally, they should be identified and resolved in parallel with the development of the innovation. In practice, however, we often see that ethics develops more slowly. An example that may be a warning is the development of commercial open access journals where both the publishing sector and the research community have failed to prevent the emergence of thousands of ethically unacceptable, predatory journals, as addressed in Sect. 16.9.3.

18.8 Variations in the Research Communities' Competence Regarding Commercialization

In the technical-industrial institutes, where R&D for the benefit of the business sector is a central part of their societal mission, managers and experienced researchers are generally well acquainted with the business community's circumstances and working methods. They are also used to assessing how the results of their own researcher-led R&D projects can best benefit society, including which results should be published and which should be patented/kept secret when this best promotes optimal and responsible use of the results. For other researchers, research results of commercial value are often a rarity. Cooperation with the business sector is then unfamiliar, and meeting with people who have other skills, operate under unfamiliar conditions, and have their own professional ethics can be difficult to handle. In very many disciplines, commercialization of the results is also completely irrelevant, and the societal benefit – in a broad sense, as discussed in Sect. 6.3 – is realized in other ways than via the business sector, and few have expertise in commercialization and cooperation with industry.

Reference

Thomson, W, 1st Baron Kelvin. (n.d.). *Wikipedia*. Retrieved September 21, 2021 from https://en.wikipedia.org/wiki/William_Thomson,_1st_Baron_Kelvin

Chapter 19
Membership of Boards, Councils and Committees

19.1 Researchers and Research Leaders Are Important Resource Persons in the Research Community

Many researchers are elected to members of peer-review panels, boards, councils and committees within the research community. The work may include handling individual cases, advising or making decisions. Many times this is a question of making assessments that can be crucial to the work and career of other researchers. This requires good professional competence, but just as important is the ability and willingness to behave properly and fairly towards other scientists. This means being accountable, open and factual and putting aside one's own interests and biases.

Researchers who agree to participating on boards and councils must also prioritize the task and set aside time for it. If they do not have time, they must decline or withdraw from the job. It does not always happen, some researchers and research leaders say 'yes' to sitting on boards, councils, editorial committees, etc., but rarely attend meetings and participate little in the work expected of them. The reason is often that other tasks must be prioritized. However, the positions of trust are still often included on the CV, where they provide a plus when applying for positions, scholarships, research funds and more. This is an expression of bad morals. Publishers and organizations who use researchers and research leaders on boards, councils and committees, in turn, also have a responsibility to ensure that the participants actively contribute, and to replace them if they do not. Having well-known researchers in such positions strengthens confidence in the activities with which the organizations are involved. If they fail to contribute, one can therefore begin to talk about deception.

© The Author(s), under exclusive license to Springer Nature Switzerland AG 2023
D. Slotfeldt-Ellingsen, *Professional Ethics for Research and Development Activities*, https://doi.org/10.1007/978-3-031-25484-0_19

19.2   Peer Review

Peer review is a case-processing method used when assessing the quality and originality of research works, primarily in connection with project applications, publication of scientific articles or contributions at conferences, job applications, evaluation of research activities, investigation of violations of research ethics norms, and more. The research community also views the method as an important prerequisite for *fair* assessment. The idea behind the method is that researchers in the same field are the most competent to assess scientific quality and originality in these contexts. Judgements by peers is also a traditional tool of justice. In recent years, peer reviews have become the subject of a more critical assessment intended to clarify the method's limitations and to understand the prerequisites for it to work well.

19.2.1   Peer Review of Scientific Articles before Publication

Today, researchers' career development largely depends on their scientific production, i.e. their (co)authorship of scientific articles. The originality and quality of their scientific production are in practice linked to whether they are published in *recognized journals with peer review* and to some extent also at conferences with a similar scheme. In public debates, it has become common for both researchers and others to try to strengthen their arguments by saying that they are based on facts and knowledge taken from peer-reviewed sources. Some also try to cast doubts on the credibility of their opponents' arguments, if they are based on something that has not been peer-reviewed. The notion that peer review is in itself a criterion for quality and fairness has established itself relatively uncontested as a mantra in the research community, to the extent that a well-known publishing editor a few years ago expressed that when something is peer-reviewed, it is in a way 'blessed' (Smith, 2006).

Gradually, however, researchers, scientific publishers and others have begun to ask critical questions about peer review as a method. Through scientific studies and more random tests, one has therefore begun to develop a better knowledge base for their use. The surveys have particularly focused on scientific publishers' use of peer review, and the results – which in part have been startling – have shed light on the shortcomings and problems with the scheme. The results, however, do *not* cast doubt on the fact that a conscientious and professionally sound peer review can significantly contribute to scientific journal articles meeting academic goals and being worthy of publication, but several weaknesses in the scheme as such have been pointed out. In the points below, some aspects of what can be seen as inherent weaknesses in peer reviews are discussed, with emphasis on the research ethics side of the issues.

- *The quality of the publishers' peer reviews is variable.* The publishers use peer review to get advice on whether or not to accept a submitted scientific article, but their quality criteria are unclear and vary considerably. The peers who evaluate a journal article are selected by the publishers and asked questions about quality and originality on the basis of the publishers' need for professional advice on this. Some journals ask more questions and have stricter requirements than others. Some are careful about who they ask for advice from, others not. Some are also better at quality-assuring peer statements than others. These variations make it difficult to look at publishers' peer review as a quality criterion in itself, and lead to uncertainty about the degree of reliability of peer reviews. To remedy this, national and international systems that group the journals according to how quality-demanding they are considered to be, have been established. Some of these are used nationally or institutionally when researchers and research institutions are rewarded on the basis of their scientific publications. The publishing industry has also addressed this; see later.

- *Many aspects of quality cannot be controlled by a normal peer review.* The peers usually only have the authors' finished document to work with. The assessment is then completely dependent on the work being truthfully and completely described by the authors. This sets limits on what peers can control. Falsifications, fabrications and plagiarism are, for example, difficult to detect without extensive investigation, access to basic data and other material (many major cases of scientific misconduct show this, but some cases of plagiarism are revealed in the peer review). Less serious scientific errors and breaches of research ethics norms may also be difficult to detect in a normal peer review based solely on the authors' presentation of their own work.

- *The peer's assessments are subjective.* Peer reviews of journal articles are normally performed by 1–3 experts who make their first assessment independently of each other. Their statements often differ, sometimes significantly. This suggests that peer reviews should be seen as subjective assessments, even though the peers have strived to be objective. In this respect, peer reviews do not differ from other assessments made by individuals. The subjective elements do not necessarily undermine a peer review as long as they seek to argue objectively and are open and honest about the grounds for their statements. On the other hand, it is an ethical problem if the inherent subjectivity of a peer review is concealed, and the use of peers is presented as an 'objective' assessment process, as is done from time to time. However, a more objective picture will often emerge when several peer reviews are compared. Some journals also allow peers to comment on each other's statements.

- *Some peers can be influenced by their own biases (assessment biases).* Some peers have one or more biases, and some allow themselves to be unduly influenced by them:

  - Methodological and scientific biases may prevent or delay the release of ground-breaking scientific methods and results. For example, it may be easier to get an article accepted if the research is within well-established methods and theories than if it takes a new path. Several Nobel prize winners have told stories that make this probable. The same most likely applies to less important research.

- Various forms of social bias can lead to certain researchers being treated unfairly. Skewed peer reviewing has been proven or made probable in the case of women versus men, researchers from developing countries versus researchers from leading research nations, unknown researchers versus known researchers, researchers at unknown research organizations versus researchers at renowned universities, etc.

The biases of peers are difficult to discover, let alone prove (many actions over a long period of time are often necessary in order to see a pattern), and bias-influenced assessments are therefore difficult to defend against. On the other hand, an author of a scientific article who is unfairly judged due to the bias of one peer will often receive fair treatment from other peers, for instance in another journal.

Bias is an 'unscientific' attribute of researchers, and those who develop a bias concerning something, stop in a way thinking analytically and objectively as researchers are supposed to do. This then leads to a certain blindness to one's own bias – it also happens to outstanding researchers. Lack of self-criticism and reflection on research ethics is therefore probably behind the development of researchers' biases, allowing biases to influence their assessments as peers.

- *Peer-review systems can violate key principles of quality assurance.* In publishers' peer-review systems, the names of the peers are usually known only to the publisher (blind process), and in some cases, the names of the authors are also kept secret from the peers (double-blind process). However, in order for the publishers' peer-review systems to be seen as an element in a credible quality assurance and control system (see Sect. 4.2), those who carry out the control cannot be anonymous. The scope of the work and the conclusions should also be openly documented so that everyone can form an opinion about how thorough the peers have been. If information about this is not available, it is difficult to know how quality-assured an article really is when it has been peer-reviewed.

- *Peer-review systems can violate key principles of transparency in research.* Anonymous peers and secret peer statements also violate research's ideal of openness. Some publishers are therefore now in favour of more openness and debate about peer statements. For example, some have started to provide the names of their peers together with the authors when the article is printed, and some also print the peers' assessments together with the article. Some online journals place less emphasis on peer review before the work is published, but allow the article to receive a tail of comments from anyone who wants to comment on it after publication.

There are arguments both against and for anonymity:
Against anonymity:

- Publication of peer reviews generally allows for a transparent, scientific discussion of the peers' statements. In practice, some adaptation will then probably be necessary.

- Most people are probably more careful and conscientious when others gain insight into their assessments. Transparency will therefore probably reduce the scope of professionally weak and ethically unacceptable work from peers.
- Openness is necessary for the publishers' peer review to be seen as an element in a quality assurance system (the names of those who review the quality, the scope of their work and their assessment must be available for the quality assurance to be credible).

For anonymity:

- It is generally more demanding to appear in public with negative rather than positive criticism. Peer reviews that are not published may therefore have a more correct balance of positive and negative criticism than statements that are published, and may thus be a better basis for publishers' decision-making. For example, it is conceivable that negative criticism of authors familiar to peers may be downplayed if the peer has to appear in public.
- For authors, open, negative criticism from peers can be stressful and probably also difficult to defend against. But in a society which is becoming more and more open, this may be something everyone is forced to accept and get used to.
- A peer who suspects that a work violates ethical guidelines may have reservations about reporting the suspicions if the statement is made public before the case has been more thoroughly investigated. Anonymous peer reviews also protect the suspect and can prevent harm and injustice if such suspicions turn out to be unfounded.*Some peers misuse confidential information for their own benefit.* It has been discovered or proved probable that peers may abuse their position to their own advantage. Dishonest peers may, for example, recommend that an article should not be published, or otherwise delay the publication, in order to have their own article in the same field published first. Others may steal ideas and results.

- *Some peers use inappropriate language.* Some peers formulate their assessments in inappropriate terms, for example with condescending, arrogant and insinuating descriptions of or opinions about the work and the authors. This is a form of bullying and abuse of power that destroys the credibility of peer reviews. Scientists must evaluate each other objectively and properly.
- *Capacity problems affect the credibility of the peer-review system.* Today the number of scientific articles is so large that publishers have problems finding enough qualified peers. This also affects the time it takes to carry out a peer review. The capacity problem has several unfortunate consequences, including:
  - Less experienced and qualified researchers, or researchers who only have more general competence in the field the article concerns, must be used as peers. The scope and thoroughness of the work must also often be reduced. Many peers, for example, limit themselves to a superficial assessment of quality and originality but check, for instance, calculations and writing. Online journals that publish thousands of articles a year and advertise that they conduct the peer review in a few weeks, can be suspected of this. An often-cited 'experiment' that sheds light on some of these journals' peer-review practices is discussed in the box below.

- Some peers take on so many assignments that it affects other work tasks. Often, agreed deadlines also have to be broken for capacity reasons. Some solve the problem by letting students help them, and some let students completely take over the task *without informing the journals*. Confidential draft articles are then also made known to unauthorised people. This is a deception that violates several norms in research ethics and undermines the trust in the peer-review system.

As can be seen from the review above, much of the criticism is rooted in the fact that some peers break with good practice for peer review. Using two or three peers to review each article may then limit the problems considerably. The possibility that all peers, independently of each other, behave incorrectly, is less likely. The precondition is that the editorial boards that receive the peer reviews take action in the event of differences in the assessment and remain vigilant to all circumstances that may raise suspicion of poor or ethically reprehensible peer review. The example in the box below shows how wrong things can go when this does not happen, and much of the criticism of the publishers' peer-review systems should probably be directed at the editors and thus at the publishers themselves. This has long been recognized in the trustworthy part of the scientific publishing industry. In 1997, for example, a number of journal editors took the initiative to establish a Committee on Publication Ethics (COPE) which advises editors, publishers, institutions and others on detecting and dealing with breaches of research ethics norms in scientific journal articles (COPE, n.d.).

An Experiment that Revealed Absent, Superficial or Unreliable Peer Review of Journal Articles – to a Significant Extent

In 2013, a microbiologist and journalist decided that he would test the quality of peer reviews in a large selection of open-access journals (Bohannon, 2013). He created a fake journal article in which he described how a molecule found in a species of lichen inhibited the growth of cancer cells. He wrote many variations of the article and thus generated a few hundred almost identical journal articles, 'written' by various, fictional authors at various fictional African research institutions. All the articles contained the same obvious and decisive methodological and explanatory errors and omissions, which a peer had to be expected to discover. Over some time he sent these articles to 304 open access journals (approximately 10 per week). Of those who were willing to review the article without advance payment, 157 ended up accepting the article, while 98 rejected it (totalling 255). The journals spent an average of 40 days to accept and 24 days to reject the article. About 60% of the 255 articles were accepted or rejected with no sign of actual peer review. For the rejected articles, this can possibly be seen as positive, if it was the case that the editors discovered that the articles were not worthy of publication and therefore did not waste resources on peer review. The most worrying, however, was that 70% of the 106 articles that went to peer review were accepted. Most peers then focused only on layout, formatting and language. Only 36 of

(continued)

the articles received peer reviews that pointed out the academic errors and shortcomings in the fake articles. Sixteen of these were still considered for publication by the editors! The experiment ended with all the articles that were accepted being withdrawn by the fake 'authors' before publication.

In a similarly startling way, several other tests of the peer-review systems conducted with fake or manipulated journal articles have shed light on how the authors' gender, institutional affiliation, etc. seem to affect the peers' assessments; how peers do not detect obvious errors; how the same article can be accepted for publication several times without it being discovered, etc. This has also been discovered in reputable journals.

Despite the problematic aspects discussed above, one must not lose sight of the broader picture – in trustworthy journals, thousands of scientific articles are reviewed each year in a professional and conscientious manner by peers around the world. This emerges in several surveys where most people think the peer reviews are useful, both for the publishers (they get a more reliable basis for accepting or rejecting an article) and for the authors (they get a quality check of their work and any advice on improvements). The peer-review system works as a filter that holds back low quality or less important research works and helps to reveal and correct some errors and ambiguities. Within the trustworthy part of the scientific publishing industry, the biggest problem is therefore perhaps not the peer-review system itself, but that without reservation researchers and others often refer to the system as a particularly reliable criterion for quality. The system's limitations and weaknesses and the number of journals with superficial peer-review procedures (see Sect. 16.9) are not mentioned equally often.

However, it is very positive that through studies, experiments and tests of the peer review system, both publishers and researchers are now searching for knowledge that can contribute to improvements. It is also positive that publishers and others are increasingly emphasizing the ethical requirements for researchers who participate in peer review.

In light of the above, peers should strive to:

- Not take on peer-review assignments without solid expertise in the field.
- Not take on more tasks than they can handle, to keep deadlines, and not seek help from others (students, colleagues) without approval from the publisher. Use the time required to be thorough and be able to answer the questions asked by the publisher.
- Not break promises of confidentiality and not misuse confidential information.
- Assess both scientific and research ethical matters, as far as possible.
- Clarify in sufficient detail what has been investigated, checked or assessed.
- Provide a factual and sufficiently comprehensive description of findings, observations and reasons for assessments.

- Be as objective as possible:
  - Have an open mind towards others' choices of research fields, theories and methods.
  - Not have prejudices related to the authors' gender, nationality, organizational affiliation, etc.
  - Not undertake peer assignments if impartiality can be brought into question.
- Use factual, objective and respectful formulations in order to:
  - Give reliable advice to the editor/publisher.
  - Give constructive feedback to the authors so that the article can be corrected or improved, and so that the critique, whether positive or negative, represents good learning.
- Take full responsibility for the review, scientifically and research ethically.

Many researchers are also involved in editorial work in scientific journals. Much of what has been said above about the professional and ethical responsibilities of peers will naturally also apply in an engagement as editor.

19.2.2   Internal Peer Review of Project Reports and Related Documents

As discussed in Sect. 4.2.2, many research organizations have procedures for internal quality assurance and control of reports from their own research projects. This is particularly necessary for the institute sector and the business sector's R&D units. Some of the peers used in the internal quality assurance will usually also have experience as peers for scientific publishers. However, the scope and content of the assignments may be somewhat different. As a peer in a publishing house, one must consider both the quality of the article and its scientific originality (at least in good journals). The latter is less relevant in internal peer reviews. On the other hand, usually a significant degree of extra work must be put into ensuring that the work maintains a high professional standard, is free from errors and ambiguities, can be ethically justified, and so on. One can also go further in proposing concrete improvements. Some of the challenges and the peculiarities of internal peer work were discussed in more detail in Sect. 4.2.2.

19.2.3   Peer Review of Project Applications

In addition to the publishers' use of peers, it is very common for research organizations and research-funding bodies (public and private) to engage peers to advise them on how project applications should be prioritized. This entails great responsibility: a rejected project-funding application has far greater consequences than a rejected publication (a less prestigious journal might often accept the article). The major research-funding bodies, such as national research councils, the European Commission, etc. have elaborate systems for this.

Review panels are often used: The peers first evaluate the applications independently of each other and then meet to come to a reasoned conclusion they can all stand behind. Each application is assessed on the basis of a group of criteria set by the funding organization, and most often the panel must rate how well each criterion is fulfilled, in addition to giving an overall score. It is important that the peers have no relations to the applicants (foreign peers are often used to ensure this). The applicants' names and organizational affiliation must then be made known to the peers, and the peers' names and assessments must be made known to the applicants afterwards. The fact that the process is open and the assessment is made by a joint panel reduces the possibility that the result will be affected by peers with preconceived opinions and biases. Those who are responsible for the application process should nevertheless pay close attention: panellists who deliberately or unconsciously break with good peer-review practice do exist.

19.2.4 Peer Review in Connection with Employment, Promotions, Etc.

In connection with employment, promotions, etc., many organizations use peers to give advice on the applicants' or candidates' qualifications. The procedures vary from organization to organization and can be both cumbersome and bureaucratic, and in part subject to statutory regulations. However, the ethical responsibility of peers is generally the same here as in other peer reviews. But when a person's career opportunities are to be decided on, the necessity of peers to act factually, responsibly and objectively becomes even more important than when evaluating project applications.

19.3 Work on Boards, Councils and Committees Within the Research System

Some researchers are elected to boards, councils and committees within research by virtue of their professional qualifications, overview, experience and positions. The tasks often concern research policy or are of a strategic or managerial nature, for example participation in compiling an R&D strategy for an organization, or for a subject area/focus area in a national or international context. The result of the work is often decisive for the research opportunities of many researchers, and several of the ethical issues related to the various forms of peer review described above also apply here.

The selection of members of boards, councils and committees is important. Depending on the group's task, three alternatives are often used in the research community:

- *The group is composed of persons who participate in the capacity of holding special positions.* The members of the group can then look at the assignment as an ex officio leadership job. Their task will usually be to promote the interests of the organization or grouping they represent and to make the necessary compromises to find a common solution the group can agree on.
- *The group is composed of representatives of special interests (for example, various research organizations or user groups), different subject areas, or the like.* The intention is that the most affected parties shall be heard and take part in decisions that affect them. The parties also gain knowledge of each other's needs and interests and an understanding of which compromises must be accepted for the sake of the whole. Affected parties who for one reason or another cannot be included in the committee must then be heard in some other way. Participation in decisions that concern oneself is considered by many to be part of a democratic process. On the other hand, there are inherent conflicts of interest in a group composed in this way, which only special circumstances can justify.
- *The group is composed of people who are expected to be neutral,* and who collectively cover the competence the group ought to have. In order to make a good assessment and a correct decision in a given case, one must generally have sufficient competence about the scientific questions in the case, and knowledge of the research environments involved. Usually this competence decreases when the distance to the subject area and the research environment increases. On the other hand, neutrality in relation to the case then usually increases. A potential weakness of 'neutral' committees is therefore that the members may lack some competence and insight into the matters they are to deal with. The case procedure should be set up in a way that helps to compensate for this.

As with the peer reviews discussed above, those responsible for the appointment of the board, committee, or council should have an open eye for the possibility that some members may have interests, prejudices and biases that they keep hidden. Measures to bring out the hidden and working methods that help the committee to think objectively, balanced and fair, can then contribute to professionally and ethically sound case processing.

Example: On request from both a major labour union and the national association of enterprises, and as an element in strengthening the competitiveness of the national manufacturing industry, the government allocates funds for a long-term R&D programme in production technology where universities, research institutes and a selection of manufacturing industries will collaborate to develop competence and technology. The national research council is given the responsibility and decides that the governmental effort must be based on a new, national R&D strategy for the subject area, prepared by representatives of the key actors: three from industry, three from the institute sector and three from the university sector. All are respected and merited professionals and leaders who together have great and broad insight into the issues. Everyone also has partial interests in the program, but the R&D strategy of a nation, organization or company should not be developed by neutral outsiders.

As in all such committees, the members show different commitments and interests: One from industry, for example, is clearly concerned that the programme should fund a considerable number of PhD scholarships – it is well-educated young people who create new jobs, she claims. No one contradicts her on that. Another of the industrialists, for his part, argues that the programme should prioritize collaboration on the development of basic robot technology that can be used in many production processes. In this, he is supported by two from the institute sector, who in addition suggest that a number of other 'cutting-edge' technologies should also be given priority. One of the professors says that it is certainly important to master modern technology, but that this can never give us lasting competitive advantages – other countries with far greater resources will always be able to beat us. Over time, however, our industry has developed a rather original and effective governance model that differs from what is common in many other countries. It contains elements that can give the national industry a distinct competitive advantage and the programme should therefore focus on further developing these opportunities, she believes. The other four committee members talk more generally but give support to and elaborate on some of the others' arguments. Some get lost in the details.

On the side line sits the research council's experienced case officer and reflects on what is being said: Everyone seems genuinely concerned with achieving a successful, national effort. All inputs are highly relevant elements in an R&D strategy for the benefit of the country. But still, it is hardly coincidental what each committee member is bringing to the table. The case officer knows, for example, that the industry leader who advocates for investing in PhD scholarships comes from a large industrial group with its own R&D laboratories where it is necessary to regularly recruit skilled researchers with a doctorate in an area relevant for the company. She also knows that the others from the industry come from smaller companies without large R&D departments, which are particularly interested in public support for collaboration with the research community. The institutes' interests in 'cutting-edge' technologies are also well known to the council. It will provide them with large, new research commissions from industry. That the professor of technology management speaks as she does is not surprising either, she is in the process of establishing a new research area within production-oriented innovation culture.

In her capacity as secretary of the strategy committee, the research council's case officer gets the committee to discuss each proposal in detail and specify who it might be useful for and how it can concretely contribute to increased industrial competitiveness. Timing issues, costs, risks, uncertainties and assumptions are also discussed. Through openness, concretization and facts, the insignificant issues are thus quickly put aside and the special interests of the committee members are largely neutralized. The chair of the committee helps to ensure that all members are active in these deliberations.

The final strategy document from the committee is then sent for consultation to the research communities and industry and, after some adjustments, is adopted by the research council. It indicates R&D areas, resource needs and forms of collaboration between research and industry that it is probable many companies will benefit greatly from, and it contains project opportunities for the universities and the institute sector that satisfy many researchers' hopes and expectations.

19.4   Impartiality

Good practice dictates that anyone who participates in facilitating the basis for a decision or participate in making a decision must not be a party to the case or be close to someone who is a party to the case. A 'party' in this context is a person to whom the decision is directed or whom the case otherwise directly affects. There must also be no special circumstances that undermine confidence in the person's impartiality.

There are generally two ways to be disqualified as a committee member:

- Automatic disqualification.

   Example: A research-funding body plans to set up a committee to evaluate applications for project funding. Persons who have a role in any of these projects or who are leaders in the organizations where the research is being carried out are considered 'partial' to the project and are therefore automatically disqualified as members of the committee. Family members too (subject to details that vary).

- Disqualification based on discretion.

   Example: When electing members of the above committee, potential members may be disqualified if there are circumstances that could undermine confidence in their impartiality. This might, for instance, be a close friendship or a professional connection to someone who is a party to one of the projects, or it could be an expressed interest in strengthening or weakening a certain type of research. Here, the degree of impartiality will depend on the circumstances and must be decided on with discretion. How the person's impartiality is *perceived from the outside* must then be given weight.

Detailed provisions on impartiality are found both in the national legislation and in the internal procedures of research organizations and research-funding bodies. The responsibility for declaring a possible conflict of interest in a case rests with the individual. The relevant provisions on impartiality are therefore mandatory reading for members of boards, committees and councils.

In order to be able to assess whether one is impartial in a case, it is necessary to have a minimum of insight into what the case concerns and who is involved in it. Apart from that, good practice dictates that persons who for one reason or another must be considered disqualified in the case, should normally not have access to case documents or be present when the case is processed and decided. The procedures for this vary by country and organization.

The provisions regarding impartiality are rooted in the research ethics' norms of openness, objectivity, fairness, accountability, reliability and more, and the formal regulations can be seen as minimum requirements to ensure that these norms are complied with.

References

Bohannon, J. (2013). Who's afraid of peer review? *Science, 342*(6154), 60–65. https://doi. org/10.1126/science.2013.342.6154.342_60

Committee on Publication Ethics. (n.d.). *About COPE*. Retrieved September 21, 2021, from https://publicationethics.org/about/our-organisation

Smith, R. (2006). Peer review: A flawed process at the heart of science and journals. *Journal of the Royal Society of Medicine, 99*(4), 178–182. https://doi.org/10.1258/jrsm.99.4.178

Chapter 20
The Researcher in the Public Arena

20.1 Knowledge Transfer and Dissemination of Research

Knowledge transfer and dissemination of research are part of the researchers' responsibility to society and go side by side with the research itself. In a research ethical context, this can often also be seen as a moral obligation to give something back to society. In practice, people expect researchers to disseminate relevant understanding, knowledge, facts, etc. accumulated in the different subject areas over time, as well as results of specific research projects. They ask for:

- Knowledge, facts, data, etc. that can be of concrete use to them.

  Example: A professor of theology writes a newspaper article about similarities and differences in the Christian and Islamic faiths and thus contributes to the knowledge base for the forming of public opinion and policy-making in a multicultural society. It is then important that the accuracy, scope, uncertainty, etc. of the information are clearly stated.

- Understanding, knowledge and facts that can entertain, delight, arouse curiosity, build competence, contribute to reflection, etc.

  Example: Based on his own research, a researcher in the history of religions writes a fascinating book about the Reformation in his home country. Not directly useful, but interesting reading. When the intention is to enrich others with stories based on accumulated scientific knowledge and research, one should be able to go further in the popularization and use a wider range of 'entertaining' tools within reasonable limits without violating norms of truth, accuracy, etc. in research ethics.

One can distinguish between:

- *Dissemination aimed at the general public.* This requires a form of expression that both arouses interest and is understood by ordinary people.
- *User-oriented dissemination.* Dissemination aimed at individuals, groups, organizations, companies and others who need concrete, research-based knowledge and facts. Users often have their own experiences and knowledge of what is

© The Author(s), under exclusive license to Springer Nature Switzerland AG 2023
D. Slotfeldt-Ellingsen, *Professional Ethics for Research and Development Activities*, https://doi.org/10.1007/978-3-031-25484-0_20

being disseminated, and the dissemination can therefore normally be at a higher level.

Therefore, all dissemination of professional knowledge and research results must be adapted to the level of competence of the target group of the dissemination. This usually requires *popularizing*. Popularization can be seen as a form of 'paraphrasing' over professional knowledge and research results, a *simplified but truthful rewriting*. All popularization thus entails a lower level of precision. In addition, it is often expected that the dissemination is engaging, captivating or exciting.

Popular, international, science-based TV documentaries fronted by researchers have helped to give research dissemination a higher status in the research community, and the forest of popular science journals, books, TV channels and mass-media features demonstrates that enriching, science-based knowledge is in great demand.

Popularization requires scientific competence, presentation skills and good discretion in research ethics. An example of how difficult popularization can be is given in the box below.

The Fossil 'Ida' – An Example of the Difficulty of Popularization
At a large-scale press conference in New York on 19 May 2009, the Norwegian palaeontologist Jørn Hurum presented a 47 million-year-old fossil of a primate, the oldest and best-preserved primate fossil ever found. The presentation was supported by a book, a TV documentary with David Attenborough and a separate website. A scientific study of the fossil was published the same day (Franzen et al., 2009). The discovery attracted great international attention, to which the form of the presentation contributed. The fossil, which was named 'Ida', then went into the collections of the Natural History Museum in Oslo, where curious people flocked to see it.

The first press release from the museum was entitled (Natural History Museum, 2009): 'THE LINK. Scientists announce the discovery of a 47 million year old primate fossil that is set to revolutionise our understanding of human evolution.' Later, wordings such as 'a link' and 'missing link' were also used about the fossil. These were obvious popularizations to create publicity about the discovery. The scientific article did not use such formulations. Researchers around the world were thrilled, but some reacted to the words used. In an article in the major Norwegian newspaper *Aftenposten* on 3 June 2009, five Norwegian professors wrote (Amundsen et al., 2009, translated from Norwegian by the author): 'To refer to Ida as a "missing link", as Hurum has done, is misleading, not informative, for the general public. Evolutionary biologists have struggled for over 100 years to overcome the delusion that we lack certain intermediaries [between humans and their origins]. We do not – on the contrary, we have found very many, and are finding more and more.' They therefore claimed that 'the fossil is presented as much more groundbreaking than it actually is, and that the research dissemination therefore

(continued)

violates basic research ethics'. However, Hurum defended his formulations and received support from other researchers, although several expressed that they themselves would use other words.

It is obvious that strictly speaking part of the presentation of Ida was not based on the research on the fossil. The popularization was not entirely true in relation to the underlying scientific material, the vocabulary a notch too 'grand'. On the other hand, Ida is a new species of primate and an exceptionally intact find that will obviously have its place in the evolutionary history of the species. That justifies some grand words. The purpose of the wording was in any case to make the discovery widely known internationally. There was a wish to disseminate exciting research and draw attention to the field of palaeontology and their own activities. Even the five critical Norwegian professors praised this: '... Hurum's engaged dissemination of fossil finds creates interest in evolution and palaeontology. This effort must simply be applauded – science needs communicators like Hurum.'

In this case, the popularization methods were thus in themselves no problem, neither scientifically nor ethically. If one thought the use of words was unacceptable, the question would be whether this was an unethical act (i.e. a violation of the research ethical norm of truthfulness and how reprehensible this was in this case) or a professional error (i.e. inaccurate popularization and how reprehensible it possibly was).

The discussion that followed in the wake of the Ida presentation thus showed that there are divided opinions about this. It should therefore be seen as an example of the grey area between what is ethically and scientifically justifiable and what is not.

Over time, a strong belief has developed that decisions in society become better when they are based on research-based knowledge. This has brought more researchers into political debates and conflicts of interest in society. For most researchers, this is an unfamiliar role that often has to be learned on the fly. Dissemination of knowledge that is to be used as a basis for decision-making in the private or public sector, or that concerns controversial issues in society, is particularly challenging. This is because research only sheds light on certain aspects of a case, and all research is associated with uncertainty. Many non-researchers find it difficult to understand these limitations. They also do not have the competence to read descriptions of research results with a critical eye, as peers do, but must be explained things in an easy-to-understand way. It then becomes particularly important to:

- Clarify both what the research results shed light on and what they do not say anything about.
- Explain the uncertainty associated with the different results.
- Give advice on how the results could possibly be used.

Knowledge and research can be disseminated in many ways through books, articles, debates, lectures, seminars and courses, TV, movies, blogs, Twitter messages, online videos, exhibitions, demonstrations, school visits, study trips, excursions and more. Many of these forms of dissemination require collaboration between the researcher and others who control the dissemination channels, such as publishers, editors, journalists and meeting organizers. Researchers therefore often have to adapt to the opportunities for dissemination that arise.

The responsibility for researchers to engage in knowledge transfer and disseminating research, lies partly with research organizations and partly with researchers themselves:

- *Research organizations* are responsible for ensuring that researchers have working conditions that promote the dissemination of knowledge and research. In particular, arrangements must be made for researchers to have enough time for dissemination, and for dissemination to be valued and career-promoting. Education and training in research communication is then an important tool. Many organizations have strategies and action plans for research dissemination.
- *Researchers* are responsible for their own dissemination within the framework set by the research organizations. The difference between the subject areas gives some researchers better opportunities to conduct research dissemination than others – an environmental researcher is much more sought after than an expert on Greek vases from antiquity. But when a palaeontologist can get people to flock to a museum to see a fossil, opportunities exist for most researchers.

Many researchers do not engage in research dissemination on a regular basis, while a few researchers are very active and visible. The latter group often stands out for their personality and commitment. Some also see a connection between significant dissemination activities and significant research activities.

20.2   Participation in the Public Debate as an Expert

Participation in the public debate is emphasized in many contexts as an important part of the researchers' duties in society. The idea is that research-based knowledge and facts can form a particularly reliable and informative basis for all debates and that factual, research-based argumentation can raise the level of a debate. This appears particularly important in the political debates prior to decisions on the structure of society and the means for it to function well.

However, the public discourse has changed radically over recent years:

- While public debates were previously dominated by certain groups in society (politicians, intellectuals, important stakeholders, etc.), the internet and the rise of social media have led to 'everyone' now speaking out. The content has also changed. Utterances based solely on subjective opinions, experiences and personal ideological, political and religious attitudes are gaining ground in the public arena. Subjective statements also seem to have a surprisingly strong impact on public debate and opinion formation.

- While the media previously gave people space or time to both highlight and argue for a case, today they are so entertaining and commercially oriented that everyone must speak to the point, short, simple, and unreserved. The PR and communication consultants' massive entry into the scenes of the public debate has contributed to this. Public debate has become a competition to sell a message. Politicians and others today are frequently quoted on what they have tweeted. Whatever can be expressed in less than 280 characters often garners the most attention.

This is the public stage that researchers encounter when they throw themselves into a public debate. A natural first reflection here is that the development has gone in a direction that should give reliable, research-based information increased value. But not everyone sees it that way. Facts have become troublesome for many, what matters is what they themselves believe, and the good feeling of expressing it. Another reflection is that researchers who, on the basis of research ethics, express themselves factually, objectively, truthfully and clearly when they participate in a public debate, increasingly encounter 'competition' from others who express themselves without such professional ethical limitations. It is not given that the researchers' communication ethics give them advantages in this competition. Unreasonable, subjective and sometimes false statements often prove to be effective. This makes the task of a fact-based and accountable expert in public debates more difficult and at times more frustrating than before, but at the same time, it seems more important than ever.

However, the benefit of researchers participating in the public debate is conditioned by the form of the debate. A debate is generally good when the participants argue objectively, exchange well-founded opinions, listen to each other, clarify facts and are open to changing views so that everyone is left with a greater understanding of the problem and better solutions when the debate subsides. However, many public debates are more marked by participants who do not really debate, but declare their own opinions, defend them when attacked, and attack others. Such debates clarify what the debaters stand for, but the interest in clarifying facts and misunderstandings and focusing on agreement instead of disagreement may be small. It contributes little to improvements in society. Research-based knowledge may fall on deaf ears in such debates, but the presence of a scientist well-trained in logic and exchange of opinions may nevertheless contribute to the debate becoming more meaningful.

20.3   The Distinction Between Professional and Private

20.3.1   *Researchers Must Clarify Whether They Are Acting as Professionals or Private Individuals*

Researchers participate in the public debate both as researchers and as private individuals. Society's special expectations for researchers to participate are linked to their *professional* contributions. In a democratic country, their contributions as private individuals will be viewed on an equal footing with of all other contributions.

Research ethics requires that researchers state when they are acting as a researcher and when they are acting privately. Under the headline Public Communication, the Singapore Statement states (World Conferences on Research Integrity, 2010):

> *Researchers should limit professional comments to their recognized expertise when engaged in public discussions about the application and competence of research findings and clearly distinguish professional comments from opinions based on personal views.* (§ 10)

National guidelines for research ethics may have similar recommendations.

An example can illustrate the importance of this:

Example: For several years, an environmental scientist specialized in oil spills has been commissioned by a large oil company to develop a mathematical model to simulate how oil spills from a production platform in the North Sea will spread in the ocean and eventually dilute and decompose. Experiments on a smaller scale have shown good agreement between model and reality. The development has taken place in collaboration with researchers in other disciplines and at several universities and research institutes, and the results have been published in scientific journals. During the planning of a new subsea oilfield, the oil company uses the model to gain knowledge about the environmental risk in the field and to plan measures to avoid damage from accidental spills. The results show that the new field, which is a little closer to land than usual, should be developed in a special way. When the oil company presents its plan publicly, referring to the environmental scientist's modelling, a marine biologist goes out in a newspaper with strong protests. He is a specialist in how environmental toxins are taken up in marine organisms and is also a strong political opponent of new oilfield developments. He first argues in general against the development of a new oilfield and then criticizes the use of the simulation model as a basis for the planning of the field. He signs the newspaper article with his full name, academic title and position. The environmental scientist responds with a newspaper article in which she corrects a number of erroneous claims made by the marine biologist related to the model and its use. She initially explains her professional background and the assignment she has had for the oil company – as good practice dictates. The answer from the marine biologist comes quickly, the environmental scientist is labelled as having a conflict of interest due to her assignment for the oil company. To the public, the marine biologist thus appears competent and trusting, while the environmental scientist appears to have been bought and paid for by the oil company. In reality, the marine biologist has no expertise regarding the spread of oil spills and the uncertainty of simulation models for oil spills. By concealing the fact that his own professional background is not particularly relevant and that he is therefore actually speaking as a politically engaged layman, the marine biologist is misleading readers, and by casting suspicion on the environmental scientist's integrity and results, he derails a knowledge-based, public debate on the matter.

Even those who do their best to separate private from professional must expect that some readers and viewers do not perceive the difference or do not wish to relate to it. For example, if a principal of a university should make a statement about something as a private person, others who react to and want to comment on the statement may find it appropriate for their own argument and case to link the statement to the institution the principal leads. That this can happen, however, is not an argument for not doing one's best to separate personal statements from statements as a professional or manager.

20.3.2   Use and Misuse of Titles

Proper use of titles and organizational affiliation when acting in public is an important tool for clarifying whether one acts as a private person or as a researcher. While noble and social titles ('Madam Director-General') are almost gone in many countries, and the importance of job titles in many professions has diminished, the use of titles in academia is still flourishing. One explanation may be that educational and professional titles such as PhD, professor, researcher, head of research, etc. still give high status and signal authority and credibility. By appearing with such titles, researchers are generally noticed, respected and listened to. As in the example above, it happens that some researchers play on this and use degree titles, job titles and organizational affiliation to appear to have authority even when they do not have special competence in the area they are writing or speaking about, or when they are speaking as a private person. For example, it is quite common for academics writing or commenting on something as a private laymen in a newspaper to sign the post with 'NN, PhD, professor at the University of …'. On the other hand, few would write 'NN, waiter, King's Arms'.

What justifies the use of titles in a modern society is that they provide relevant information about the person who uses them. If the information is not relevant to the case, the titles are unnecessary – they are social snobbery.

> Example: A professor of labour law writes an article in a newspaper about harassment in the workplace and states that he is 'PhD John Johnson, professor at the University of Oxford'. Most readers will then get the impression that the author is very well qualified in the field – an assumption which in this case implies correctness. If the writer had not been a law professor with expertise in harassment, but a professor of mathematics, then most people would probably think that the use of the title is misleading. The mathematician may have wise views, but his academic background and professorship are not relevant to what he writes about. The use of the title would make many people perceive him as professionally competent in the field of harassment, which he was not. Misleading people is ethically unacceptable.

Many research organizations will have internal procedures for the use of job titles and organizational affiliation. The principles will normally be:

- When speaking on behalf of a research organization, a project group or similar, it is good practice to state the academic title, position title and organizational affiliation. The purpose is to account for the formal competence to speak on behalf of an organizational unit or the like.

- When speaking in the area of one's own expertise, it is good practice to state one's professional title, job title, subject area and organizational affiliation. The purpose is to state the formal competence and responsibilities.
- When speaking as a private person (i.e. not on behalf of a research organization or the like, nor as an expert), it is good practice not to state the academic title, job title and organizational affiliation.

The responsibility for sending the right signals lies with the individual researcher and leader. However, newspaper editors, journalists, interviewers, moderators and others are often concerned with presenting people as 'impressively' as possible. Even if, for example, no titles or organizational affiliation have been given in a private newspaper article, the newspaper editorial staff may use headlines such as 'Harvard professor expresses ...'. Experienced researchers are prepared for this and make sure to specify their competence in relation to the case, for example by opening the newspaper article with: 'I do not have relevant professional expertise in this case and do not speak on behalf of my research organization but...'. If, in a TV interview, conference or the like, one is presented as professor in a context where the professorial expertise is not relevant, one might start by saying that: 'You introduced me as a professor, but I want to emphasize that I am not expert in the field we are now talking about...'.

20.3.3 The Danger of Being Politicized

Researchers who either work in politically disputed areas or who are themselves politically active in areas where they do not have special professional knowledge, sometimes express how difficult it is to keep politics and professional work separate. This may happen in several contexts, for example:

- Researchers who speak out in public with expert knowledge relevant to a political issue can be interpreted as speaking politically and be associated with a party in the conflict.

  Example: A researcher has for many years studied how the national culture is being influenced partly by multicultural impulses from abroad, partly from within through resident immigrants. She believes the research results give her special competence to participate in the immigration debate, and starts with a newspaper article where she presents objective facts about the changes in national values, ways of thinking and traditions, without expressing that anything is positive or negative. However, the political wing in favour of immigration sees this as an expression of her negativity towards the change and loss of national values, and is quick to accuse her of being a nationalist. Thus, she is in practice 'parked'. In order for her professional contribution to the public debate to be heard, she must now first make people understand that she does not belong to the nationalist wing of politics and that she is concerned about facts, not politics. In an area of the politics full of emotions, prejudice and political games, this can be difficult. Repeated over a long period of time, this can damage researchers' reputation and credibility.

- Researchers who privately show great political involvement in a case may also be ascribed political motives in other contexts. One can in a way be politicized as a professional and be exposed to suspicions and accusations of not being factual and objective as a researcher. Researchers who sign political petitions or make regular appearances in the public debate are particularly vulnerable to this.

Example: A researcher in the history of the Middle East is not particularly politically engaged, but after an attack on a synagogue, he comes in empathy to sign a petition expressing sympathy for those attacked. Palestine-friendly debaters interpret this as an expression that he is pro-Israel in all contexts, and pigeonhole him with other political opponents. When he speaks, they counter him with unsubstantiated allegations that are apt to question the reliability of the research he is conducting on the Israel-Palestine conflict, and to cast suspicion of bias on him. Defending himself against this with objective and factual arguments can be difficult.

For *leaders* in research organizations, this can be particularly problematic because the organization is often identified with its leaders. For example, a leader who as a private person signs a political petition may make people believe that the organization supports the cause, especially if the leader signs with full title and organizational affiliation. (If the governing bodies of the organization have decided that the organization should support the petition, this is of course fine.) This can also be problematic internally in the organization, where there may be employees who do not support the cause.

Confidence in the objectivity of research is undermined when researchers, leaders and research organizations take a political stand, and the clearer one expresses a stand, the greater the expected reactions must be. However, this must be weighed against the right of everyone to express a political opinion.

Generally, society has a high level of confidence in researchers. A survey in several countries worldwide showed, for example, that scientists were rated as the most trustworthy of a number of professionals, followed by doctors and then teachers (Ipsos, 2019). Despite the high ranking, only 60% in this survey considered scientists to be trustworthy, while 11% considered them to be untrustworthy to varying degrees. In another global survey, 18% of the respondents who expressed an opinion had low trust in scientists (Gallup, 2019, chart 3.1). The lack of trust may, for example, be due to suspicions that researchers allow themselves to be controlled or influenced by others (authorities, clients, etc.), and that researchers allow their personal opinions and points of view (political, ideological, etc.) to influence their work. As an example of the latter, a survey among the population in Norway showed that while about 8 out of 10 had strong or somewhat strong trust in *research*, about 3 out of 10 of the respondents totally agreed or somewhat agreed that 'research results are to a large degree influenced by the researchers' own political attitudes and perceptions' (Kantar, 2019, p. 10 and 26).

Most researchers likely believe that the ideal of neutrality is well observed in the research community. However, the surveys mentioned above indicate that there is probably a gap between this self-image and ordinary people's perception of researchers. This should provoke reflection and action in the research community: How can one prevent researchers from inadvertently or intentionally politicizing their research? How can one avoid falling undeservedly suspect of being politicized?

The ideal of truth is so strong in research that the vast majority do their best to be objective. But some may allow their own political interests to influence both the content of the work and the interpretation of the results – many times probably without noticing it. Politicization can also occur when researchers disseminate research-based knowledge and participate in public debates – and perhaps first and foremost then. Those who follow the public debate closely will not have any difficulty in discovering researchers who in articles, interviews and debates popularize or argue in a way that goes a little too far in the direction of the unscientific, in a way that makes their personal political attitudes and views shine through. Often it is a matter of the small things – perhaps just a few words – but enough to flag that the researcher is not completely neutral and thus can easily come under suspicion – rightly or wrongly – of letting personal political interests influence the research. Although there are obviously degrees of politicization, and it is unreasonable to demand that researchers be completely unaffected by their own political views, the point here is that *it takes little to lose people's trust.*

Some researchers may see it as a societal responsibility or duty to contribute to the use of specific scientific knowledge in some areas in society, and believe that this both necessitates and justifies that they enter the political arena and use tools that to a certain extent can break with the ideals of objectivity and neutrality in research. Researchers, like everyone else, must be allowed to participate in political debates. At a time when many are questioning researchers' political and ideological neutrality, it is at the same time important that researchers act in a way that does not undermine but rather strengthens confidence in research.

20.4   Marketing, PR or Lobbying Activities Within Research

Marketing, PR and lobbying are extensive activities in society, primarily in politics, business and interest groups, but also in other areas where there is competition for positions and funds, such as in art and culture, sports, and research. This is nothing new – these activities can be traced back to ancient societies.

Most research organizations and researchers are in constant competition with each other for research funding. Everyone, therefore, has an inherent interest in positioning themselves, their organization, their subject area or their project in an advantageous way in relation to their competitors. Some put this into action. Most researchers choose to create a buzz about what they are fighting for using conventional means: writing an article in a newspaper or magazine, arranging a seminar, giving lectures, being interviewed in a newspaper, radio or TV, demonstrating something exciting in a laboratory, during fieldwork or similar. This has a lot in common with the ordinary dissemination of research and scientific knowledge, but the purpose is different. Others are more direct and ask for a meeting with the decision-makers to present their case. In all marketing and lobbying, timing is very important. It's about giving decision makers and opinion leaders a story that excites and feels right, just at the right moment.

In the context of research ethics, it is difficult to see anything fundamentally wrong in research organizations and researchers promoting their own interests in different ways, as long as it takes place in a responsible manner. Here, the requirements for accountability and truthfulness should be the same as for the dissemination of knowledge and research otherwise. If the purpose of marketing is to improve the opportunities for project funding, research equipment, etc. one must not, for example, fail to mention both positive and negative aspects. In a world full of misleading marketing, this can be difficult. The particularly strong position of research in society depends entirely on people's trust. It disappears easily if research organizations and researchers fall for the temptation to oversell or conceal negative factors.

Example: A research group wants a new technology area to be prioritized within an industry-oriented, thematic programme in the national research council. Shortly before the programme committee is to meet, the group arranges a seminar where the field's exciting opportunities are highlighted. Key people who can influence the content of the programme are among the participants and note the interest in the field among those present. The researchers will thereby ensure that the subject area's scientific potential and societal importance are well known and understood both by some of the decision-makers (ministry, research council, programme committee) and among other opinion makers. Of course, they also hope that some of the decision-makers will be particularly enthusiastic about the subject area and be left with the feeling that it is both right and safe to invest in it. The presentations at the seminar sheds light on the exciting research opportunities the new technology area allows and how the results are likely to benefit industry and society in general. None of what the researchers say is wrong, and their enthusiasm can in no way be seen as overselling. What they fail to mention, however, is that any applications of the technology will probably be 10–20 years into the future and that no national companies as of today have the technology to directly utilize any research results. They also fail to mention that there is an alternative technology, just as exciting and potent, at the forefront of which are several research groups abroad. The audience therefore leaves the seminar inspired and with the feeling of being well informed about the new technology area, but unaware of many key factors that are important when planning such a research programme for the benefit of domestic industry.

The temptation to oversell the importance of one's own research and projects is driven by competition for research funding. The researchers feel a need not to tell the whole truth in order to secure the opportunity to carry out the research they are passionate about. The research-funding bodies can help prevent this, among other things by ensuring that the processes associated with the use of research funds are perceived as open and fair.

References

Amundsen, T., Folstad, I., Giske, J., Slagsvold, T. & Stenseth, N. C. (2009, June 3). 'Ida' er over-solgt. ['Ida' is oversold. In Norwegian]. *Aftenposten*, Kultur, 4.

Franzen, J. L., Gingerich, P. D., Habersetzer, J., Hurum, J. H., von Koenigswald, W., & Smith, B. H. (2009). Complete primate skeleton from the middle eocene of Messel in Germany: Morphology and paleobiology. *PLoS One, 4*(5), Article e5723. https://doi.org/10.1371/journal.pone.0005723

Gallup. (2019). *Wellcome global monitor – First wave findings.* London. Wellcome. Retrieved September 21, 2021, from https://wellcome.org/sites/default/files/wellcome-global-monitor-2018.pdf

Ipsos. (2019). *Global trust in professions. Who do global citizens trust?* Ipsos. Retrieved September 21, 2021, from https://www.ipsos.com/sites/default/files/ct/news/documents/2019-09/global-trust-in-professions-trust-worthiness-index-2019.pdf

Kantar. (2019). *Befolkningens tillit til forskning og syn på Forskningsrådet.* [The population's trust in research and views on the Research Council. In Norwegian. Report to The Research Council of Norway]. Oslo. Kantar. Retrieved September 21, 2021, from https://www.forskningsradet.no/siteassets/publikasjoner/2019/befolkningens-tillit-til-forskning-og-syn-pa-forskningsradet

Natural History Museum. (2009, April 19). *The Link* [Press release]. Retrieved September 21, 2021, from http://www.revealingthelink.com/more-about-ida/resources/press_release.pdf

World Conferences on Research Integrity. (2010). *The Singapore statement of research integrity.* Retrieved September 21, 2021 from https://wcrif.org/statement

Part III
Violations of Research Ethics Norms:
Extent, Handling and Reactions

Chapter 21
Extent of Violations of Research Ethics Norms

Breaches of responsible research practice can occur in all phases of the research work, be it in planning, implementation, reporting, peer review, etc. However, such breaches are first and foremost detected when they manifest themselves in documents to which others gain access. Research has traditionally had a reputation for being accurate, truthful, honest, verifiable, neutral, legal, etc. This impression has occasionally been stained by sensational revelations of misconduct that have gained international media coverage. Until recently, these violations of research ethics norms have been seen as anomalies and few have questioned whether research is really as 'clean' as one would like to think. However, as research has increased in scope and societal importance, there has been a need to more closely investigate the extent to which the prerequisites for society's particular confidence in research can be said to be valid (see box below). The results of these first surveys were startlingly negative. They indicated that deviations from good research practice could be said to be common in the research community. Here, of course, less serious cases will dominate, but according to several surveys, in the order of 1% of researchers have probably at some point committed serious violations of good research practice, such as falsification or fabrication! An overview of some major misconduct cases of international concern and some studies of the extent of hidden misconduct can be found in Anderson et al. (2013).

As an example, the first surveys in Norway (Bekkelund et al., 1995; Elgesem et al., 1997) indicated that 5% of those who answered the questions had knowledge of falsification and fabrication of data at their own faculty or department. This caused shock and debate. The results, which were somewhat uncertain, were met with denial in many places in the research community, but also became a national wake-up call that there might be something wrong with our attitudes to research ethics. In a large survey of researchers in Norway in 2018 (Hjellbrekke et al., 2018), just over 2% of respondents stated that they knew that colleagues had participated in falsifying data or material (about 8000 researchers in full or part-time research positions answered the survey). For fabrication and plagiarism, the corresponding

D. Slotfeldt-Ellingsen, *Professional Ethics for Research and Development Activities*, https://doi.org/10.1007/978-3-031-25484-0_21

numbers were just over 2% and just under 14%, respectively. The survey also included a small selection of less serious breaches of good research practice. Here, approximately 40% of the respondents stated that they had broken with good practice one or more times during the last 3 years (Hjellbrekke, et al., 2019). For example, about 10% had given co-authorship to people who did not meet the criteria for co-authorship, and nearly 30% had knowledge of colleagues who had done so. The corresponding figures for denying co-authorship to persons who were entitled to it were 2 and 13%, respectively. Just over 40% stated that they were not familiar with the procedures for reporting suspected scientific misconduct, and approximately 5% stated that they themselves had failed to report breaches of ethical standards in the last 3 years. In a large survey in the Netherlands, about 8% admitted to having falsified or fabricated their research at least once in the past 3 years, while about 50% admitted that they had frequently committed at least one violation of responsible research practice (for details on the scope of the survey and its uncertainties; see Gopalakrishna (2021)). These surveys indicate that rather many researchers are aware that others around them secretly violate research ethics norms and that surprisingly many do so themselves.

Examples of Surveys on the Extent of Scientific Misconduct
Through a relatively small number of surveys and analyses published around 2010, the international research community was suddenly made aware that deviations from responsible research practice are not uncommon, and that research ethics must therefore be given greater attention in all research organizations. The examples below are excerpts from four of these studies.

1. Fang et al. (2012) reviewed scientific publications within a limited subject area that had been withdrawn for one reason or another. The study included the article database PubMed, which contained 25 million biomedical articles from the mid-1940s to 2012. The first of these articles to be withdrawn was published in 1973 and withdrawn in 1977. Between then and until 3 May 2012, 2047 articles had been withdrawn. Only 21.3% of these were recalled due to errors. 44.7% were withdrawn due to falsification or fabrication (or suspicion thereof), 14.2% due to duplicate publication and 9.8% due to plagiarism. The percentage of articles withdrawn for reasons other than errors has increased since 1990. As several research groups accounted for many withdrawals, the number of research groups involved in these cases is smaller than the figures indicate. Although the percentage of withdrawn articles is low, the numbers reveal weaknesses in research's quality assurance systems. The average period from publication to withdrawal was over 30 months, and several of the withdrawn articles were frequently cited, even after they were withdrawn. In cases of frequent withdrawals, the cause was not given or was reported in a dishonest way – which in itself must be regarded as serious misinformation. The results were dominated by withdrawals from reputable journals.

(continued)

2. Martinson et al. (2005) conducted an anonymous survey among U.S. researchers who were in the early or middle stages of their careers and who had received support from the National Institutes of Health in the United States. The researchers were asked to answer 'yes' or 'no' to whether, during a specified three-year period, they themselves had broken with one or more of 16 specific forms of unacceptable research practice of varying severity. Of the 3245 researchers who responded, 33% stated that they had breached at least one of the forms of unacceptable practice. Some examples:

   – 0.3% had falsified or 'cooked' research data.
   – 2.4% had used other people's ideas without permission and without crediting sources.
   – 6.0% had refrained from mentioning data that contradicted their own previous work.
   – 15.5% had changed design, method or results in a study in response to pressure from funding sources.

3. Titus et al. (2008) asked a couple of thousand researchers who had received support from the National Institutes of Health in the United States if they had seen cases of falsification, fabrication or plagiarism in their research department in the last 3 years. Those who responded stated that they had observed approximately 200 cases of misconduct, corresponding to three cases per year per 100 researchers.

4. Fanelli (2009) conducted the first meta-analysis of investigations into the extent of fabrication, falsification and modification of data. He found 18 studies that could be included in the analysis. He found that on average 2% of the researchers in anonymised surveys stated that they themselves had at least once falsified, fabricated or modified data, and 14% stated that they knew others personally who had done so. Meanwhile, 34% stated that they themselves had also committed other questionable research practices. The analysis showed a significant range in the figures, but the main conclusion is difficult to ignore. Most of the studies included in the meta-analysis were in medicine/biomedicine in the United States.

There have been several more studies following these, and many more will likely appear in the years to come. What is most important about them is that they have so far proved that unacceptable research practice is not a rarity. At the same time, it is probably wrong to paint the situation black (Fanelli, 2018), as some seem to do. When evaluating research results, the starting point should be that the results are reliable. At the same time, one must keep in mind that there will always be someone who does not follow good research practice – or simply cheats.

References

Anderson, M. S., Shaw, M. A., Steneck, N. H., Konkle, E., & Kamata, T. (2013). Research integrity and misconduct in the academic profession. In M. B. Paulsen (Ed.), *Higher education: Handbook of theory and research* (pp. 235–241). Springer.

Bekkelund, S. I., Hegstad, A. K., & Førde, O. H. (1995). Uredelighet i medisinsk og helsefaglig forskning. [Misconduct in medical and health research. In Norwegian]. *Tidsskrift for Den norske legeforening, 25*, 3148–3151.

Elgesem, D., Jåsund K. K., & Kaiser, M. (1997). *Fusk i forskning. En studie av uredelig og diskutabel forskning ved norske universiteter.* [Cheating in research. A study of research misconduct and questionable research at Norwegian universities. In Norwegian]. Oslo. The Norwegian National Research Ethics Committees. Retrieved September 21, 2021, from https://www.nb.no/nbsok/nb/e67b07808d7b871478c78da8090c74fc?lang=no#0

Fanelli, D. (2009). How many scientists fabricate and falsify research? A systematic review and meta-analysis of survey data. *PLoS One, 4*(5), e5738. https://doi.org/10.1371/journal.pone.0005738

Fanelli, D. (2018). Opinion: Is science really facing a reproducibility crisis, and do we need it to? *Proceedings of the National Academy of Sciences (PNAS), 115*(11), 2628–2631. https://doi.org/10.1073/pnas.1708272114

Fang, F. C., Steen, R. G., & Casadevall, A. (2012). Misconduct accounts for the majority of retracted scientific publications. *Proceedings of the National Academy of Sciences (PNAS), 109*(42), 17028–17033. https://doi.org/10.1073/pnas.1212247109

Hjellbrekke, J., Drivdal, L., Ingierd, H., Rekdal, O. B., Skramstad, H., Torp, I. S., & Kaiser, M. (2018). *Etikk og integritet i forskning – Resultater fra en landsomfattende undersøkelse.* [Ethics and integrity in research – Results from a nationwide survey. 1st sub-report from the working group in the research project RINO (Research integrity in Norway). In Norwegian]. Norway. University of Bergen, The Norwegian National Research Ethics Committees and Western Norway University of Applied Sciences. Retrieved September 21, 2021, from https://www.uib.no/sites/w3.uib.no/files/attachments/rino_delrapport_1_2018_2.pdf

Hjellbrekke, J., Ingierd, H., & Kaiser, M. (2019). *Diskutabel forskningspraksis: Holdninger og handlinger.* [Questionable research practices: attitudes and actions. 2nd sub-report from the working group in the research project RINO (Research integrity in Norway). In Norwegian]. Norway. University of Bergen, The Norwegian National Research Ethics Committees and Western Norway University of Applied Sciences. Retrieved September 21, 2021, from https://www.uib.no/sites/w3.uib.no/files/attachments/rino_delrapport_2_2019.pdf

Gopalakrishna, G., ter Riet, G., Cruyff, M., Vink, G., Stoop, I., Wicherts, J. M., & Bouter, L. M. (2021). *Prevalence of questionable research practices, research misconduct and their potential explanatory factors: A survey among academic researchers in The Netherlands.* MetaArXiv Preprints. Retrieved September 21, 2021, from https://osf.io/preprints/metaarxiv/vk9yt/

Martinson, B. C., Anderson, M. S., & de Vries, R. (2005). Scientists behaving badly. *Nature, 435*, 737–738. https://doi.org/10.1038/435737a

Titus, S. L., Wells, J. A., & Rhoades, L. J. (2008). Repairing research integrity. *Nature, 453*, 980–982. https://doi.org/10.1038/453980a

Chapter 22
Various Forms of Irresponsible Research Practice

22.1 General

The research organizations are responsible for reacting to and investigating irresponsible research practices committed by their own employees. In order to make correct use of their resources, the different forms of irresponsible practice should be sorted into categories according to type and severity. Sloppiness must be treated differently than cheating. Many research organizations will probably find it suitable to distinguish between four categories of ethically irresponsible behaviour that in principle should be handled differently: *sloppiness and negligence, less serious violations of research ethics norms, serious violations of research ethics norms (research misconduct)* and *breaking the law.* The first thing to do every time a new case arises is to obtain an overview of its facts and circumstances, so that it can be placed in one of these categories. Case officers must use their best judgement here. If a case turns out to be more or less serious than first thought, adjustments can be made along the way. The purpose is to process the cases at the right level in the organization and use sufficient, but not excessive resources, while complying with research ethics guidelines, the organization's internal procedures, and national legislation (more on this in Sect. 23.5). The four categories are discussed in more detail in the following chapters.

22.2 Sloppiness and Negligence

Here, 'sloppiness and negligence' implies deviations from good practice due to sloppy work and inadequate quality assurance, i.e. first and foremost a breach of the expectation of accuracy in research. Sloppiness and negligence can be viewed as low-quality academic work that should not be downplayed or overlooked. It must be

D. Slotfeldt-Ellingsen, *Professional Ethics for Research and Development
Activities*, https://doi.org/10.1007/978-3-031-25484-0_22

prevented, stopped and reacted to in the work environment around the individual or individuals concerned.

All scientists make mistakes. Most are detected by the researchers themselves or by different quality assurance measures, but some may remain undetected for a long time. Making such 'honest' mistakes is not immoral. However, when researchers have little regard for the quality of their own work and do not do their best to ensure that everything is in order – then the term 'sloppiness or negligence' is justified. This manifests itself in many ways, for example:

- Deficiencies and inaccuracies in obtaining and registering research data.
- Errors or inaccuracies in descriptions of methods and procedures, data, calculations, statistics, etc.
- Errors in marking quotations and paraphrases and in referencing other sources.
- Quotations that are inaccurately reproduced, and paraphrases that do not quite match the source.
- Illogically arranged and/or vaguely worded research reports.
- Errors and ambiguities regarding facts in applications, CVs, etc.
- Errors or incompleteness in test logs (for example, forgetting to keep a log one day).

Sloppiness and negligence rarely cause great harm to anything or anyone, but it does happen. Aside from saving time, sloppiness and negligence are probably likely to be actions that rarely serve a purpose or an intention.

Sloppiness and negligence occur particularly often when research work is reported, and then far more often in less prestigious project reports and notes than in scientific journal articles, where the authors usually put a lot of work into being accurate. This is especially bad when the individual who is sloppy also has an inability to properly formulate a text. Reports can become so inaccurate, ambiguous and incomplete that the reader is seriously misled about what has been done and the results of the research. Good research practice dictates that all research reports, even the less prestigious ones, should be as accurate and clear as scientific journal articles. In research organizations where there is a large production of reports that are not published in scientific journals, one should therefore have a special focus on accuracy and clarity of the reports and consider establishing extra measures to prevent sloppiness.

22.3   Less Serious Violations of Research Ethics Norms

'Less serious' violations of recognized research ethics norms implies actions that on the one hand cannot be excused as sloppiness and negligence, and on the other hand are not so serious that they fall into the category of research misconduct. Traditionally, this type of deviation from good research practice has been termed 'questionable' research practice (QRP). Questionable research practice is not, however, a precise term because in most cases this will actually concern actions that most researchers find unacceptable. There may, though, be different opinions about how serious some of these actions are, and where the boundaries should be drawn between what is acceptable and what is not. The danger of using the term 'questionable' is also that many may consider themselves entitled to interpret the research ethics guidelines in their own way. Therefore, some argue for using the term 'undesirable' research practice here, or alternatively 'unacceptable' practice (used in ALLEA (2017), and elsewhere).

The bullet points below give examples of violations of research ethics norms that the research community has generally regarded as less serious:

- Giving co-authorship to persons who do not meet the criteria for co-authorship, or failing to list persons as co-authors when they meet these criteria.
- Splitting up the research results and publishing them separately to get more publications out of the work ('salami publication') or, related to this, to publish the same material (or essentially the same) several times (for example in a journal and at one or more conferences with proceedings) without giving notice ('duplicate publication').
- Failing to update information, such as listing rejected articles as 'in press' in one's CV or withholding information about research work that fails or is delayed, and, if caught doing so, calling it an 'oversight'.
- Writing too positive or too negative attestations or letters of recommendation.
- Behaving poorly towards colleagues and students, including unreasonable accusations against others.
- Letting an assistant carry out a peer-review assignment without reporting it.
- Failing to report breaches of recognized research ethics norms.

More examples can be found in ALLEA (2017).

The first four examples include various forms of violation of the norms of truthfulness in research ethics. Carried out intentionally or in gross negligence, some of these may be considered as scientific misconduct (for example, if someone who has contributed to a research project and clearly meets the criteria for co-authorship is knowingly and willingly omitted as a co-author of the publication by colleagues). If the research ethics guidelines are enforced consistently and uniformly, nationally and internationally and over a long period of time, one can envisage that violations of norms that are currently considered less serious may in the future be regarded as more serious. In other words, the boundary is subject to discretion and is not immutable.

22.4 Serious Violations of Research Ethics Norms (Research Misconduct)

22.4.1 The Definition of Research Misconduct

The definition of research misconduct varies somewhat by country and organization. The core of the term is (ALLEA, 2017):

Research misconduct is traditionally defined as fabrication, falsification, or plagiarism (the so-called FFP categorisation) in proposing, performing, or reviewing research, or in reporting research results:

- Fabrication is making up results and recording them as if they were real.
- Falsification is manipulating research materials, equipment or processes or changing, omitting or suppressing data or results without justification.
- Plagiarism is using other people's work and ideas without giving proper credit to the original source, thus violating the rights of the original author(s) to their intellectual outputs. (From 3.1 Research Misconduct and other Unacceptable Practices)

The core of the international perception of research misconduct is thus various forms of violation of the central norms of truth in research ethics, primarily fabrication, falsification, and plagiarism. However, some also include other serious violations of research ethics norms in the term. In everyday discourse, research misconduct that is committed intentionally may often be referred to as cheating, deception or fraud. Cheaters always have a motive for their actions. It may be to make the research results more important or publishable; gain recognition as a researcher; save time; earn money from commercialization of research results; downplay or make others' contributions to the research invisible; hide illegalities in the research; hide the fact that paid work was never performed; promote their own professional, ideological, political or religious opinions; secure funds for their own research; improve their chances of being offered a scholarship or position, etc.

In practice the three main forms of research misconduct will cover many different types of breaches of good research practice, as described in the following.

22.4.2   Fabrication – Cheating that Cannot Be Explained Away

Fabrication is making up data and other results and recording or reporting them as genuine research findings. Fabrication is also a form of falsification. Some of the most extensive cases of dishonesty that have come to light have included fabrication. The holder of the 'world record' in withdrawal of scientific articles, at least up to 2012, was an anaesthesiologist who was forced to withdraw a total of 183 articles featuring fabricated data. He had never seen the patients he reported in his clinical trials ('Yoshitaka Fuji', n.d.). The most common motive for this form of research misconduct is probably to achieve status and reputation. Fabrication can be difficult to reveal but is at the same time very difficult to explain away when it is first discovered.

22.4.3   Falsification – Also when Not Telling the Whole Truth

Research materials, research methods, results, reports, etc. can be falsified in many ways. Examples are:

- Alterations to data and other results.
- Selective removal of unwanted data or other results.
- Selective application of methods and procedures to achieve a desired result.
- Withheld information about the research work that may be important for the assessment of the work.
- Misrepresented entry of data and information obtained from other sources.
- Misrepresented interpretation of results and conclusions.
- Manipulated research log.
- Misleading information in project applications, job applications, CVs, etc.

One important point here is that it can be just as wrong not to reveal the *whole* truth as to say something directly untrue. This is often a challenge when writing a research report. All research projects generate more data and information than is possible or appropriate to report. For example, the space provided by publishers in scientific journals is limited, and some provide only space for the most original and important results. Within commissioned research, a client can request that the focus of the report is on useful results. The challenge then is to be both concise and to present the full truth about the research work. In this process, discretion must be applied, and discretionary assessments can always be debated and criticized. This can lead to conflicts and accusations of falsification. An example can illustrate this:

> Example: A researcher conducts a meta-study of changes in ocean temperatures globally. He then finds data of highly variable accuracy and reliability. After evaluating the data, he chooses to ignore a few rather uncertain measurements that differ from the others. He finds this so obvious that, primarily for reasons of space, he chooses not to mention and discuss the omissions when the results are published in a journal article. Another researcher discovers this and accuses the researcher of falsification, in that data that contradicts the study's conclusion have been omitted.

> The question of who is right here can only be determined by studying the details of the case. On the one hand, it is conceivable that the author's professional assessments are found to be acceptable and that the accusation of falsification is unjustified (perhaps the accusation was even made to harm the author, as a result of a conflict between the parties). Maybe it was just a matter of the researcher showing poor craftsmanship. On the other hand, it is conceivable that the author's assessments were not tenable, that he should have explained the omission of the data in more detail, that there is reason to suspect him of having been influenced by a predetermined view of the temperature changes in the ocean, and that the term falsification may therefore be fitting.

It is important to be aware that a cheater can view omissions of data and information as less 'dangerous' than direct manipulation of facts because if they are detected omissions can in some cases be rationalized. The cheater may, for example, point out that the selection of what is included and discussed in a report or article is the result of a professional discretionary assessment, that the assessment may be wrong, but that this does not imply falsification. Sound scientific expertise must be engaged both to reveal such explanations and to possibly confirm that the explanation is acceptable.

Not publishing a research result when, for some reason, it is not favourable to the researcher's interests, or to the interests of others who influence or have power over the researcher, is also related to this. Examples can be:

Example 1: A study of a pedagogical method used in schools shows results that undermine the researcher's personal political views on school systems. The researcher completely refrains from publishing the study.

Example 2: A study of a medical treatment shows it has a lesser effect than that demonstrated in previous studies. The researcher, who has invested great professional prestige in the treatment concept, completely refrains from publishing the study.

Example 3: A drug study performed on behalf of a pharmaceutical company shows some unexpected side effects. The client does not benefit from this becoming public and leans on the researcher not to publish the work.

When research works are not published, however, it is probably most often because they are not completed or do not meet goals in terms of quality. Some works that researchers seek to publish are also stopped by publishers (ethical issues related to the publishers' assessments were addressed in Sect. 16.9.5).

22.4.4   Plagiarism – The Most Common Form of a Serious Breach of Good Research Practice

All research projects are based on the use of results and information from previous projects, both one's own and those of others. From a research ethics point of view, the precondition is that it is clearly stated what has been obtained from others and where the source is. Despite the fact that it is easy to do this correctly, each year some researchers are caught using text, figures, data, etc. from other sources without being sufficiently clear about the origin. Many of them are found guilty of plagiarism.

The term 'plagiarism' is not usually used about minor matters, but about serious cases and usually under particular circumstances (see later). In most cases, plagiarism is a *deliberate* act.[1] For example, those who cut and paste from others are well

---

[1] Some believe that the term plagiarism should be used only when it is committed intentionally. Others *also* include gross negligent acts in the term plagiarism. The latter is used in this book.

aware of what they are doing. Some refer to plagiarism as a form of *theft*, but in the context of research, plagiarism is more about a *hidden use* of material from other sources whereby the reader is *misled* about who has done what, and where the material originally comes from. In particular, plagiarism is a violation of three important basic principles in research ethics: truthfulness, traceability and respect for the contributions of others. The latter is connected with the fact that the efforts of those who are being plagiarized are *made invisible*. Being referred to by other researchers is meritorious, and the number of citations is an important parameter when measuring the scientific significance of the work of individuals, groups, organizations and countries. Anyone who uses something from the work of others without referring to the source is undermining this system and taking credit from research colleagues. This seriously violates the norms of how researchers should treat each other.

Plagiarism of Text

The most common form of plagiarism in research is plagiarism of text. It includes:

- *Quotations* that are not marked as quotations and where no reference is made to the source. The reader is then not made aware that both the wording and content of the text originate from others.
- *Paraphrases* of text from other sources where it is not indicated that it is a paraphrase, and where no reference is made to the source. The reader is then not made aware that the content of the text originates from others.

In research ethics, these types of breaches of good practice is usually termed plagiarism if the author intentionally misled the reader, was aware that it would mislead the reader, or should have understood that the action could have such an effect (legally this means that the author has acted with degrees of intent or gross negligence).

One also often finds examples of breaches of the good practice of a related but less serious nature:

- The *quotation* is not marked, but a reference is given in connection with the unmarked quotation that refers to the source. The reader can then not know that this is a direct quotation, and may often think that there is instead a poorly marked paraphrase. The reader is thus made aware that something is taken from elsewhere, but will be in doubt as to how much of the text refers to the given source.
- The *paraphrase* is not marked as a paraphrase, but a reference is given in connection with the unmarked paraphrase that refers to the source. The reader is thereby informed that there is reference to another source but will be unsure of what this is.

This is considered a breach of good practice that must be reacted upon, but will often not be seen as a serious violation or plagiarism if its extent is not too severe.

Surveys in several countries indicate that text plagiarism may be quite wide-ranging. It is conceivable that this has much to do with the use of modern word processing programmes and the availability of documents on the internet that enable extensive use of cut and paste. Combined with poor work routines in front of the PC, there are apparently many who lose track. However, electronic document processing has functions that make it possible to keep track of what one has retrieved from other sources, so it is really just a matter of making use of these possibilities and maintaining effective habits when using a computer. Fortunately, however, the new technology that makes it easier to be sloppy and cheat also affords us new opportunities to expose those who plagiarize. Publishers and research institutions are making increasing use of plagiarism-checking software to routinely investigate whether a submitted journal article, dissertation or the like contains text sections that can be found in other documents on the internet. This contributes to some plagiarism being revealed at an early stage.

'Self-Plagiarism'
Related to plagiarism is what is somewhat illogically termed 'self-plagiarism'. Self-plagiarism is reusing something (text, figures, tables, etc.) from one's own previous works without referencing it as such. This must not be confused with cases where, for example, two researchers, A and B, do not mark a quotation from a previous article that A and C have written. Then it is a matter of plagiarism in the normal sense because B takes part of the credit for something he or she has had no part in, while C's contribution is made invisible. It is not much help if A claims to be the main author and has written most of both texts if the reader is misled about the co-author's contribution.

Another form of self-plagiarism is 'duplicate publishing', characterized by the same work being published two or more times (possibly in different versions and with slightly different titles) *without disclosing it*. Here it does not matter if one version or issue is published in a scientific journal while another is presented at a conference. By issuing duplicate publications, the author appears to be more productive than they really are. However, publishing the same, or almost the same, two or more places when the author *clearly states* that it is a duplicate publication, is ethically acceptable (no one is misled), provided the duplication is also indicated in publication lists and CVs.

Self-plagiarism is less serious than plagiarism because other people's work is not being presented as the author's own. It is still, though, a violation of research ethics norms because the author is withholding the fact that certain elements stem from a previous work (information about the chronology of research results is important). Self-plagiarism is therefore a violation of the principles of truthfulness and traceability. There are different perceptions about the severity of the different forms of self-plagiarism. Serious cases can be considered research misconduct.

Other Forms of Plagiarism

Most plagiarism cases involve text and figures. However, plagiarism can also be the use of others' ideas, theories, interpretations, designs, illustrations, etc. without giving appropriate reference to the source. Many such cases concern disagreements about the use of project ideas and project plans. Two examples were given in Sect. 8.1; a third is given below:

> Example: A researcher in a research institute gets the idea for a technological research project and develops the scientific and administrative foundation for the project. The idea is based on a fairly original combination of her own professional knowledge and ideas about possible applications. She addresses an industrial company that she believes will be interested in the project and proposes collaborating on an assignment basis. She wants the company to finance the project. In return it will get the exclusive right to use the results within its area of business. The company finds the project very interesting but so risky that it prefers to start with a smaller preliminary project. In addition, it wants a well-known foreign university, which has special expertise in the field, to participate. The researcher agrees that this would be useful. The preliminary project is successful, and the company is ready to move on, but now wants a PhD student at the foreign university, supervised by a professor there, to conduct the work. The university has the competence to do so and in addition some equipment that is not available at the research institute. It also costs less for the company to pay for the PhD scholarship than for the commissioned work at the institute and it is not so important that the PhD student will need more time. The company thanks the researcher and pays her institute for the work she has done with the preliminary project. In the contract with the company, the researcher and her institute have not taken into account the possibility of this happening. They feel manipulated by the company, but decide to leave it there; they would rather spend time on good clients than on unfaithful ones. The project is then modified to suit a PhD study but otherwise continues mainly according to the original plan. When the project is finished the student publishes the parts of the results that have no direct commercial interest to the company in an international, scientific journal, as agreed with the company. He then refers in an acknowledgement to an initial preliminary project, thanks the company for financial support and the professor for guidance, but does not mention the institute researcher anywhere. The company, in turn, uses the results in its further development projects.

> Although the company has paid for the time the institute researcher worked on the preliminary project, it has not paid anything for her idea and preparatory work, which turned out to lead to commercially interesting results for the company. For the researcher and the institute, this would have been fine if she had been given the work on the main project, as the plan originally was. Instead, the company 'stole' the project idea and gave the job to someone else. Since the managers of the industrial company are not researchers, they cannot reasonably be judged according to guidelines for research ethics. Whether the company's actions can be defended must therefore be assessed on the basis of laws, agreements and ethical norms that prevail in the business world; this is beyond the scope of this book. On the other hand, the actions of the PhD student and the foreign university that took over the project and reported the results without stating that the idea originated with others must be assessed on the basis of international research ethics norms. It is then quite obvious that the idea for the project, in this case, came from the researcher, and that the PhD student should have stated this in his thesis. However, in this case, as in all cases, the details must be carefully reviewed before concluding that plagiarism has taken place. This example also shows how important it is to document one's own ideas and secure the rights to the ideas when entering into collaboration with others.

Plagiarism in Relation to Copyright
The term 'plagiarism' is also often used in connection with infringement of the copyright of intellectual works. However, the criteria for infringement of copyright are not the same as the criteria for breaches of good research practice described above (although there are similarities). This was discussed in greater detail in Chap. 17.

Different Degrees of Severity in Plagiarism
Plagiarizing one sentence is wrong, but is generally considered less serious, while one page will usually be considered a serious violation. However, the boundary between the serious and the less serious is not clear, and there are differing views in the research community as to where this boundary should go. Two examples of cut and paste that went wrong can be indicative for what is considered plagiarism by some, *at least*; see the box below.

Two Examples of Researchers Who Were Found Guilty of Plagiarism
A professor together with several co-authors submitted an article to a well-known scientific journal. The journal's peer-reviewers discovered that 15 lines in the introduction (including some references) were taken verbatim from another work. The lines were not marked as quotations and the reference was missing. The professor apologized and explained that it was his fault. He had cut and pasted the lines from the other source and intended to process them further, but later forgot that they had been retrieved from elsewhere (the work on the manuscript had been going on for some time). The journal described this as plagiarism, the work was refused publication in any form, and the professor was banned from publishing in the journal for 3 years. The university where the professor worked issued him a reprimand. The co-authors were exonerated.

A PhD student submitted a thesis in which it was discovered that eight sections of text (seven from the introduction) were hidden quotations from other sources (in total under two pages of a thesis of over 200 pages). The student laid his cards on the table, said he had not intended to cheat, and explained that he had lost track of what he had taken from other sources. Towards the end of the work he had become stressed by the shortage of time. The university's research ethics committee found that this was plagiarism, and thus research misconduct. The doctoral thesis was rejected.

The two cases were investigated locally at two Norwegian universities and reported to the National Commission for the Investigation of Research Misconduct (the commission's internal documents, 2012).

In the borderland between what everyone would consider serious and less serious plagiarism, respectively, we often see that comparable plagiarism is judged somewhat differently by the various organizations or committees investigating the cases. There is also a somewhat broad view of what can be regarded as mitigating circumstances in plagiarism. For example, some emphasize that researchers have an independent responsibility to familiarize themselves with good practice for using material from other sources (everyone starts learning about this at an early stage in school). Others believe that a lack of teaching and guidance in research ethics at university or workplace, colleagues with bad habits, etc. may be mitigating circumstances that can make it wrong to use the term 'research misconduct' even if there is an objective case of plagiarism.

To avoid ending up in this grey area, everyone should follow the guidelines for referencing quotations and paraphrases and carefully refer to the source regardless of the scope and content of what is taken from elsewhere. To speculate that anything else may be acceptable is to jeopardize one's own research future. Similar caution should be exercised with regard to other forms of plagiarism (plagiarism of ideas, etc.).

General Differences in Severity Between Fabrication, Falsification, and Plagiarism

Essentially, fabrication, falsification, and plagiarism are always considered serious breaches of good research practice. However, the degree of severity will depend on what has been done wrong, its extent and the damage it can cause. Many also believe that there is a certain inherent difference in severity between fabrication and falsification on the one hand and plagiarism on the other. The main reason for this view is that the consequences of plagiarism are usually less damaging than the consequences of fabrication and falsification. Fabrication or falsification contains untruths, while plagiarism usually contains truths but misleads in terms of what has been done by whom. Plagiarism is therefore primarily detrimental to the people who are plagiarized, while fabrications and falsifications can cause harm far beyond the research community.

Another reason why many take fabrication and falsification particularly seriously is that these actions are difficult to detect, while plagiarism will eventually be revealed – sooner or later. For preventive reasons, it is therefore important to have *zero-tolerance for fabrication and falsification*. Many would also argue that a researcher who does not refrain from any form of fabrication and falsification completely lacks the moral attitudes required to conduct research. They also always know what they are doing and cannot claim that they do not know it is wrong. Everyone also knows that fabricated or falsified research results are almost worthless. It is therefore difficult to find any mitigating circumstances in such actions.

Although the research community does not have the same zero-tolerance for plagiarism the many individual cases that appear in the media show that plagiarism can cause as great a breach of *trust* between the researcher and society as fabrication and falsification. In 2011, the German Minister of Defence had to resign because someone discovered that in his doctoral thesis from the University of Bayreuth in

2008 he had cut and pasted text from other sources without specifying where the text came from. In 2012, it was the President of Hungary who had to resign due to plagiarism in a doctoral thesis from Semmelweis University in Budapest in 1992, and in 2013 the Minister of Education in Germany had to resign due to extensive plagiarism in a doctoral thesis from Heinrich Heine University in Düsseldorf in 1980. All three stated that they had not intended to cheat, but all were nevertheless deprived of their doctorate and had to leave their positions of trust. The research community and society in general thus reacted strongly to plagiarism. The course of events in these examples also illustrates that the plagiarist must expect to be found out – and it is only a matter of time before it happens.

22.4.5   The Question of Whether One Can Be Held Responsible for Research Misconduct in Unfinished Works and Documents

Fabrication and falsification can take place both during the research work and during the writing of scientific articles, theses, project reports etc. These types of serious breach of good research practice must be considered committed at the moment they are carried out, and both basic data, work journals and other draft documentation that shows that this has taken place, in addition to completed documents and data, can be used as a basis for investigations and conclusions regarding research misconduct. In some cases, one may find that a researcher has carried out falsifications or fabrications during the research work, but has later come to their senses and corrected the wrongdoings in the final documents that describe the work. It cannot be assumed that this is entirely absolving, but perhaps mitigating, and each case should be considered on its own merits.

In the case of *text plagiarism*, the situation is different. The correct marking of quotations and paraphrases, and complete and correct references, are often completed only in the final phase of the work on a manuscript. Co-authors, partners, managers and supervisors, clients and others who may read and comment on the draft document in the final phase, also often provide input and views that influence which sources one should mention and include material from, and thus who to refer to. It follows that it must be wrong to hold someone responsible for apparent 'plagiarism' in documents that have not been completed by the authors. Investigations and conclusions in these cases must therefore in practice be based on articles that the authors have sent to publishers or websites for publication, dissertations sent for evaluation, final assignment reports sent for approval to the organization's management or to a client, a research-funding body, etc. However, in order to protect oneself from being accused of plagiarism in unfinished documents (it does happen), it may be wise to state on the document that it is unfinished and that the references are not final.

In the case of *other forms of plagiarism* (plagiarism of ideas, etc.), it is difficult to say anything in general about when in the work the misconduct occurs; the circumstances of each case will be decisive.

22.4.6 Fake Science

In recent years, fake science has become a phenomenon that must also be addressed. Fake science is an activity that pretends to be serious research, but which is not, and which can contain both falsifications, fabrications and plagiarism. Fake science is often produced by non-scientists, most often to make money on falsification or to make a product, a treatment, a political position, etc. far more credible than it really is by lying that it is founded on research. Researchers everywhere are regularly enticed to take part in this, for example by being tricked into collaborating with untrustworthy researchers and others; to co-author junk articles; to join editorial committees in predatory journals, etc. (the names of well-known and trustworthy researchers help to camouflage the fraud). To agree to take part is to contribute to fraud. Until now, the research community and society at large have often looked at researchers who have been lured to this as innocent and unsuspecting victims of deception. However, researchers have a duty to understand what they are agreeing to participate in. If they become involved in fake science, they should therefore be investigated on an equal footing with everyone else who is suspected of scientific misconduct.

22.5 Violations of Laws, Regulations, Procedures and Agreements

Failure to comply with laws, regulations, procedures and agreements is ethically unacceptable research practice that undermines confidence in research. These cases range from violations of national criminal codes (for example, in the case of embezzlement or fraud in connection with research funds, corruption in connection with commissioned research, harassment of colleagues, etc.) and personal data protection laws (in research that includes the collection and handling of personal data), to breaches of terms in a research-funding contract. Such cases often have both legal and ethical aspects. They therefore constitute a separate category of unacceptable practice that usually has to be dealt with in special ways. The primary responsibility for this lies with the research organization, but in many cases the police, public supervisory authorities, etc. will also become involved. Within the research organization, such cases are probably best dealt with at the top level, possibly with lawyers as experts and advisers when relevant. Special research ethics matters related to these cases may be investigated by an internal ethics commission or similar, in addition to or as a basis for the assessment of the legal aspects of the case. The latter is beyond the scope of this book; the former is discussed further in Chap. 23.

References

All European Academies. (2017). *The European code of conduct for research integrity* (revised ed). ALLEA. Retrieved September 21, 2021, from https://allea.org/code-of-conduct/
Yoshitaka Fujii. (n.d.). *Wikipedia*. Retrieved September 21, 2021, from https://en.wikipedia.org/wiki/Yoshitaka Fujii

Chapter 23
Handling of Violations of Research Ethics Norms

23.1 The Duty of Research Organizations to Respond to Violations of Research Ethics Norms

All employers have an interest in and responsibility for their own employees basing their work on recognized professional ethical standards, and for taking action if an employee fails to do so. If someone is suspected of having violated research ethics norms, it is, therefore, *the research organization* that first and foremost investigates the suspicion and sanctions those who have acted wrongfully. Each research organization will have *internal procedures* for this (see, for instance, ALLEA (2017, Sect. 3.2)). In essence, these procedures should specify:

- How the research organization will process external and internal whistleblower reports (also anonymous) where suspicions or allegations of breaches of recognized research ethics norms are made.
- How the whistleblowers' interests and rights will be safeguarded.
- How the interests and rights of suspects and others who may be parties to the case are safeguarded.
- How the research organization will process and make a decision in the case.
- How the research organization will react if a breach of research ethics norms is found (including how to sanction the person who has done wrong, how to inform others about it, how to correct or limit the damage, etc.).

National laws may have provisions on how this must be carried out.

Based on general legal principles and research ethics norms, one can derive a number of general expectations for the research organizations' internal procedures in this context; see the box below.

D. Slotfeldt-Ellingsen, *Professional Ethics for Research and Development Activities*, https://doi.org/10.1007/978-3-031-25484-0_23

Overall Expectations of Research Organizations' Handling of Suspicions of Violations of Research Ethics Norms

- The procedures for processing cases must be familiar and accessible to all (expectation of uniformity and transparency).
- The leaders of the research organization must respond to all deviations from responsible research practice (expectation of accountability and equal treatment).
- In serious cases, an independent and competent ethics commission or the like should state an opinion on which violations of norms exist, and on the degree of guilt of the persons involved (expectation of equal treatment, objectivity and impartiality).
- Cases must be processed quickly and without undue delay, but accurately, objectively and thoroughly (expectation of consideration and accountability).
- If there are multiple suspects, the complicity and guilt of each suspect must be assessed individually (expectation of justice).
- Suspects must be treated as innocent until proven otherwise (expectation of fairness and consideration).
- Whistleblowers must be treated properly (expectation of respect).
- In order to protect whistleblowers, suspects and others involved, the case should be treated confidentially and exempted from public access until a conclusion is reached (expectation of fairness and consideration).
- The suspects and others who are parties to the case shall have full access to the case, be given the opportunity to explain themselves, respond to or comment on others' allegations, case submissions and case presentations and present their own evidence (expectation of everyone being heard)
- The research organization must take actions against anyone found guilty of having violated recognized research ethics norms. The reaction must be adapted to the severity of the violation (expectation of justice).
- The research organization must be adequately open about the cases, learn from them and use them for improvement (expectation of openness and quality).

(The list is based on a discretionary selection of general legal and ethical principles. ALLEA (2017) is particularly concerned with much of this).

23.2   General Aspects of the Procedure for Dealing with Violations of Research Ethics Norms

The procedures that the research organizations follow when investigating suspicions of violations of research ethics norms are influenced by the relevant legislation, and some elements of case processing may also be directly regulated by law.

This differs from country to country. This also applies to part of the terminology related to the investigation of breaches of ethics. It is therefore not possible to give an exact description that is 'valid' in all countries. In the following, therefore, only a review of the principal steps of such an investigation is given, using layman's terminology. The readers must familiarize themselves with the details and formalities for how this is done in their own country and organization.

In all cases involving suspicions of breaches of recognized research ethics norms – serious or less serious – there will be three main aspects of the case that must be clarified and assessed.

1. First, it must be clarified (evidence must be found) whether in fact there has been a breach of research ethics norms, i.e. a breach of established practice or custom in the research community, or a breach of written institutional, national or international guidelines for research ethics. In the following, this will be called the 'objective element' of the blameworthy act (in line with the legal term 'actus reus', guilty act, used in some countries).
2. Secondly, it must be clarified (evidence must be found) whether the person who has broken with good practice can be blamed for it. In what follows, this will be called the 'subjective element' of the blameworthy act (in line with the legal term 'mens rea', guilty mind, used in some countries).
3. Once these facts have been clarified, the following must be decided upon:

   – Whether any appropriate sanctions should be imposed on guilty parties. The sanctions must comply with the legal and institutional regulations. This applies to all violations of research ethics norms, not only research misconduct.
   – Whether any appropriate measures should be taken to correct errors (such as withdrawing a scientific article), limit any damage, improve routines and practices, etc.

23.2.1 The Objective Element of a Blameworthy Act

The objective element of the suspect's actions include:

- What the suspect has specifically done that violates recognized research ethics norms (for example: what data has been fabricated, or which text elements are unmarked quotations).
- Which specific common practice, ethical guideline, internal procedure, etc. has the suspect violated (for example: the Vancouver recommendation of co-authorship).

When assessing the severity of the objective elements, there are usually three things to consider:

- *How seriously the research community in general views the type of breach of good research practice that has been committed* (for example, it is considered far more serious to fabricate data than to list a co-author who should not have been there).

- *The extent* of the breach (for example, it is far more serious to fabricate 50 observations than one).
- *The consequences* of the breach (a falsified research work that can lead to incorrect treatment of patients is, for example, far more serious than a falsification that has no bearing on anything or anyone).

Normally, the seriousness of a case will in principle be related to the action itself, while its scope and consequences contribute to increasing or reducing its severity. In this lies a discretionary element that one sees many examples of when comparing cases processed by different organizations and in different countries. With regard to falsification and fabrication, however, many (as discussed in Sect. 22.4) will consider *the act itself* so serious that there are few mitigating circumstances. Such cases are always serious.

23.2.2   The Subjective Element of a Blameworthy Act

The subjective element of the suspect's actions has to do with the degree of culpability. It is then a question of the person's state of mind and the circumstances surrounding the action at the time of the wrongdoing. For example: Was it the person's intention to cheat? Did he knew that his actions were wrong? Should he have known it? Was he misled into acting as he did?

Here, too, there will be different degrees of culpability (for example, it is more blameworthy to intentionally hide a quotation than to forget to mark the same quotation as a result of poor work routines). National legislation and national or institutional ethical guidelines may have provisions on how the degree of culpability is to be assessed. On a general basis, it may be appropriate to distinguish between:

- *(Simple) negligence* (the researcher should have known that the action broke with good research practice, but the violation is less pronounced).
- *Gross negligence* (the researcher should have known that the action broke with good research practice, and the violation is pronounced and highly blameworthy).
- *Intention* (the researcher knew that the action broke with good research practice and was aware that it could have consequences).

The question of the degree of culpability must be answered on the basis of *facts* concerning the circumstances surrounding the act.

In some cases, a researcher may have seriously violated research ethics norms (the objective aspects of the act are proven), but that special circumstances in the case mean that the subjective conditions for guilt are not present. One may then be faced with a situation that cannot be described as research misconduct, but which can nevertheless be highly reprehensible.

Example: A group of researchers at a research institute has, on a commissioned basis, worked for several years to strengthen the general knowledge base within a technology area of interest to a large, industrial company. The results are communicated to the client in a series of project reports. The authors vary. On one occasion, the group's youngest researchers writes a report on her first work for the client. To put this in context with the group's previous work and for the sake of simplicity, she chooses to extract texts from some previous reports, written by others in the team around her. The text from previous reports, a total of ten pages, is inserted almost verbatim, with minor adjustments. Neither quotation marks nor direct reference shows that the text comes from elsewhere, but in the list of references, she includes all the reports the group has written for the client. The report is verified by a senior researcher in the group and approved by the department head. A researcher at the client discovers that there is a comprehensive, unmarked transcript from previous reports. He suspects research misconduct and reports to the head of the institute about this. The institute's research ethical committee, which is investigating the case, has no doubt that the objective conditions for plagiarism have been met. On the question of subjective guilt, however, they are more uncertain. The researcher explains that she did not intend to mislead anyone and that she assumed that the client would see and understand what were new and what were old results. She also states that neither the experienced researcher who quality assured the report, nor the head of the department who approved it, reacted to the lack of citation marking and reference, even though on reading they must have understood that previous results were repeated for the sake of context. The research ethical committee takes this into account in addition to the researcher's inexperience. They weigh this against researchers' independent responsibility for following good research practice. They conclude that the facts related to the question of guilt in this case do not indicate that there is a strong probability that the researcher has actually behaved in a way that can be described as scientific misconduct. However, they criticize the researcher strongly for incorrect citation practice and at the same time criticize the institute for poorly implemented quality assurance. The institute's top management follows this up by issuing a written warning to the researcher and tightening up the internal quality assurance routines. Another ethical committee might have come to a different conclusion; assessments of guilt often vary.

23.2.3 Assessment of Uncertain Evidence

When there is a suspicion of a violation of research ethics norms, one must, as indicated above, start by finding evidence that proves the actual circumstances associated with the act. Some evidence is more uncertain than others.

The objective elements of the action can then often be substantiated by relatively certain evidence. For example, a long, unmarked text that is identical to a text in another, older document, is 100% certain to be an unmarked quotation. Reported research data related to a patient who has never lived, is with 100% certainty fabricated data. However, if a laboratory assistant testifies that a researcher has reported results of a chemical analysis that are different from what the assistant measured, and that original data has been accidentally lost, then it is not 100% certain that falsification has taken place. Here, the explanations and credibility of the assistant and the researcher must be assessed against each other, and one may end up with an uncertain assumption that the assistant should be given a little more weight than the researcher, illustrated by quantifying the probability of falsification to 50–70%. In most cases, there will be much evidence to deal with. The different elements of evidence must be weighed against each other to reach a reasoned conclusion, and the uncertainty of the conclusion must be stated and justified.

Corresponding assessments must be made with regard to the subjective elements of the act, the question of guilt. Here, however, the evidence is almost always uncertain because it consists of information about the researcher's motives, thoughts and assessments, competence, position, the situation in the immediate research environment, etc. Often one has to rely largely on written and oral testimonies from those involved and from witnesses. Testimonies are always laced with uncertainty, but when supported directly or indirectly by documented facts, the overall level of uncertainty can be reduced. In the same way as when the objective elements of actions are assessed, one must also here weigh different evidence against each other, the conclusion must be substantiated, and the uncertainty of it must be stated.

Uncertainties are sometimes quantified as a percentage probability that a specified event has occurred ('there is a 70–80% probability that NN has fabricated data'). Others find it more meaningful to put the uncertainty into words ('there is clear and convincing evidence that NN has fabricated data').

Because it is difficult to state uncertainty accurately, it is appropriate only to distinguish between a few levels of uncertainty, for instance:

- *Preponderance of evidence* that the act actually took place – i.e. that there is between 50% and approximately 70–80% probability that the act actually took place.
- *Clear and convincing evidence* that the action actually took place – i.e. that there is upwards of 70–80% and close to 100% probability that the action actually took place.
- *Beyond any reasonable doubt* that the action actually took place – i.e. that there is a close to 100% probability that the action actually took place.

In some countries the terms and definitions used here as examples are based on the law. These categories vary from country to country and there are different opinions about where the boundaries between them should go in terms of percentage of uncertainty.

When investigating serious violations of research ethics norms, many will think that the evidence must at least be clear and convincing in order to find a suspect guilty. National legislation or institutional procedures may, however, have provisions concerning the requirements of evidence related to violations of research ethics norms.

23.3 Open or Confidential Investigation of Violations of Research Ethics Norms

In all investigations of suspicions of breaches of responsible research practice, one will have to decide on a number of questions related to the openness of the case: Should the public have access to the case? Should the investigating organization actively inform the public about it? Should the names of those involved be made known? When and how should this possibly happen? In the context of research ethics, this is a question of weighing two important principles against each other: the principle that research should be open and accessible, and the principle that everyone involved in the case should be treated fairly.

The public can get access to a case in two ways:

- By law, individuals have the right to access certain documents.

  - National legislation has provisions on this, at least with regard to public research institutions. The provisions are often intricate and detailed, and readers should therefore always familiarize themselves with relevant national laws on this point. Persons who are a party to a case have special rights with regard to access to documents relating to their own case. In this chapter, however, only the general right of access is discussed.
  - Few make use of these rights. It is often only when journalists happen to become aware that a case is under investigation, request access to documents, and write about the case, that it becomes more widely known.

- The research organization may actively inform the public about the case.

  - Usually, neither public nor private research organizations have any statutory duty to actively make a case public. For research ethics or other reasons, however, they may choose to inform the public about it, or about parts of it. If so, there are many forms and degrees of active publication; see next section.

23.3.1 Research Ethics Assessment of Confidentiality and Transparency in the Serious Cases

While legislation imposes different requirements on public and private companies with regard to access to cases concerning breaches of research ethics norms, there is no difference between private and public research organizations when the degree of transparency in such cases is assessed in the light of research ethics. Some of the main issues related to this are addressed below.

Ethical Reasons for the Active Publication of Serious Violations of Research Ethics Norms

Truthfulness, honesty, etc. are key elements in the moral contract of researchers and research organizations with society. If someone breaks with their professional ethical obligations, society, therefore, has a reasonable 'right' to know about it. All researchers also use the work of others and trust that good research practice has been followed. If this is not the case, the research community has a corresponding 'right' to know about it. These are arguments for transparency about violations of research ethics norms, at least in serious cases. On the other hand, it is difficult to find arguments that everyone is entitled to know about *suspicions* of violations of research ethics norms. Exceptions may be cases where for one reason or another the case and the suspicion must be made public at an early stage.

All organizations and individuals who become involved in cases that include violations of research ethics norms have a significant responsibility to be careful when informing others about the case. The information should be truthful, accurate, fair and objective, and the informant must take responsibility for the consequences of revealing the information about the case in the way chosen. For example, a research organization that, in the name of transparency, immediately or without due diligence goes public with a whistleblower's suspicion of research misconduct, might in practice contribute to the spreading of untruths if it turns out that there is no basis for the suspicion, or that the allegation has been made with malicious intent.

Consideration for Fair Treatment for Suspects, Whistleblowers and Others

The need for society and the research community to become aware of serious violations of research ethics norms is not the only aspect that must be given weight when the research organization dealing with the case is to make a decision on taking the case public. Consideration for those directly involved in the case must also be taken into account. Everyone must receive fair treatment.

This applies in particular to the suspect(s). In many cases, it turns out that the suspicions are unfounded or that the case is far less serious than first thought. If a case becomes publicly known while there is still only a suspicion (for example based on reports or allegations from a whistleblower), there is a risk that speculations rather than facts may affect collegial discussions about the case and that inaccuracies and exaggerations may occur in media coverage of it. The suspect is likely to have difficulty defending him or herself against this as long as the case is under investigation. The risk is that the person in question may experience loss of credibility, difficulties in collegial relationships and problems in conducting research work while the investigation is ongoing. In the most serious cases, the suspect's immediate family and close colleagues may also be affected. This violates the principle that one should be treated as innocent until proven otherwise and is not, therefore, fair treatment of the suspect.

Consideration must also be given to a possible whistleblower's situation. If the whistleblower's name becomes publicly known while the case is under investigation, there is a risk that he or she may be exposed to difficulties in collegial matters and in other ways, for example by questions being raised about the whistleblower's motives and dealings with facts.

This suggests that the case should in principle be treated confidentially until all matters concerning it have been clarified. This should be the normal according to ALLEA (2017, Sect. 3.2) 'in order to protect those involved in the investigation'. Since research is an international collaboration to a large extent, it is appropriate to follow international guidelines in this area. National legislation and institutional procedures may, however, contain provisions on this. Confidentiality implies that information about the case is only given to persons who need to know about it along with a requirement that the information continues to be treated confidentially. This will in the first instance mainly include the persons involved in the actual processing of the case, relevant leaders in the research organization, all who are defined as a party in the case, and their representatives and advisers.

There may also be other reasons why a case should not be made public while it is being processed. Most commonly, this concerns cases where accusations of breaches of good research practice are made by one or more parties in a collegial conflict. This, then, will not only be a question of assessing whether someone has broken with good research practice, but also of resolving the conflict. The latter can often be the most important aspect and the core of it all. Personnel matters should be resolved 'behind closed doors', where it is easier for the parties to admit mistakes and make compromises, and where the parties' statements about each other are least harmful. Ideally, however, one should always try to separate the assessment of the alleged breaches with good research practice from the work of resolving the collegial conflict. In those cases where the two issues have to be treated together, confidentiality will almost always increase the chances of finding workable solutions.

However, there may exceptionally be reasons why some information about the case should actively be made public during its processing. One such reason may be that more or less reliable information and rumours about the case have become made public through newspapers, TV, social media, etc. as a result of leaks from people with some knowledge of the case. This may make it necessary for the research organization dealing with the case to go out with correct information earlier than planned.

In some cases, the interests of a suspect must also give way to other considerations. This applies primarily when the suspicion of a breach of research ethics norms includes matters that must be dealt with as a precaution to limit or avoid any possible harm or damage immediately and before all the facts have been obtained. If, for example, there is a suspicion that research results have been falsified, and that human health and safety might be in danger, it would be irresponsible not to immediately inform those who could prevent harm from occurring. This may result in the case receiving public attention. If the suspicion proves to be unjustified, every effort must be made to limit the harm and burden that the suspect and others may have suffered.

When a suspicion of a breach of responsible research practice has been investigated and a conclusion has been reached in the case, some of the arguments for confidentiality dissipate. The suspect is then either acquitted of the suspicion or found guilty of wrongdoing. In the latter case, the severity and degree of culpability have also been assessed. Information about the case can then be based on facts and well-considered assessments. This is therefore the right time to weigh the consideration of the person who has acted wrongly against the consideration of society's and the research community's need to be informed about irresponsible research practice.

In both public and private research organizations, there may be several arguments for actively going public with the case. For instance:

1. That it is necessary for the sake of the general trust in researchers and the research organization.
2. That it is necessary to correct errors, such as withdrawing or correcting a publication or report.
3. That it is necessary to prevent or reduce the harmful effects of what has been done wrong.
4. That the case can be used as an example to other researchers and society that violations of research ethics norms are not accepted, that control is exercised, and that measures are taken against those who act wrongfully.
5. That the case contains ethical boundaries or matters of principle that should be made known.
6. That the acquittal of persons who have been wrongfully suspected of wrongdoing shall be publicly known.

In all these cases, the content of the information should be assessed on the basis of what others need or have a 'right' to know about the case. In case 1 above, the most important thing may be to show that the research organization is keeping its own house in order, while the name of the person who has acted wrongly is often not so important. In some cases, however, it may be important to reveal the name of the person who has acted irresponsibly. In case 2, it can often be argued that the research community and others are entitled to an explanation of why an article or report has been withdrawn, and knowledge of the authors' role in this (there is a significant difference between withdrawing something due to honest errors and scientific issues, and to withdraw something due to research misconduct). If the co-authors have varying degrees of guilt in what has been done wrong, this must come to light. In case 3 above, it may be sufficient to inform those directly affected by the case. In cases 4 and 5, there is rarely a basis for identifying the persons involved when the case is made public. In case 6, it will be appropriate to publish the information about the case that contributes to making the suspect free of blame. In the assessment here, it is natural that the opinions of the person to whom this applies are given special weight. When such reasons for active publication are not present or are not so important, the research organization may be best served with a passive form of transparency, i.e. to provide information about the case upon request.

Whenever active publication of a case is justified, there will always be questions about how this should be done: Send out a press release or post information on social media? Publish the entire investigation report (possibly anonymised), or just a brief abstract of the case? The answers here will vary from case to case, but there is one issue that must be given special consideration. This applies to information that is posted on or may find its way to the internet, where one must expect that it will remain for years. If, for example, one 'googles' the name of the person who in 2006 committed the most severe research misconduct discovered in Norway, thousands of findings come up, and the screen is covered with pictures of him. A person who has done something really wrong must live with the fact that many people will know about it for a long time. It is far worse to live with the fact that the case is constantly exposed in the world's most accessible public arena and appears every time someone searches for the same surname on the internet. Even though at the time the case was serious on an international scale, it now seems unfair that it should pursue the person in question and all his close relatives so openly, probably for the foreseeable future. It is difficult to see the internet's exposure here as anything other than a modern form of the pillory – which has no place in modern society. Based on the principle of fair treatment of those who have been found guilty of wrongdoings and consideration for their close relatives, one should therefore try to avoid or limit the extent to which people are identified by *name* on the internet (the name may in some way be exposed on the internet, but the fewer times it is exposed, the less damage and burden it can cause). As far as possible, one can also establish a routine for removing old cases from the internet when they have lost their relevance.

Although most people involved in a case of research misconduct probably sense the public awareness of the case as a burden, the openness can of course in certain cases also be of benefit to (some of) those involved. A suspect who is acquitted may, for example, wish to make this outcome public knowledge, especially if it has already come to light that the person in question has been a suspect. The same can apply to whistleblowers who have their suspicions confirmed, especially if they have experienced problematic reactions from colleagues, managers, clients or others as a result of the whistleblowing.

In summary, if the research organization finds important reasons to inform the public about a serious breach of research ethics norms, the ideals of both openness and fairness in research ethics can often be realized if the research organization issues *a brief statement* regarding:

- What the case is about, with facts about what has been done wrong.
- That the research organization has imposed sanctions against the person who has acted irresponsibly and that the sanctions are in accordance with the research organization's internal procedures (only in very special cases can there be reasons to be specific on the sanctions).
- How the case has been handled (here it can often be enough to refer to internal procedures).
- The names of those who have handled the case (the names of the decision-maker, case manager, the members of an eventual ethics commission or other experts who have been consulted).
- What has been done to correct errors and limit any damage.
- Whether the case has revealed errors or deficiencies in the research organization's internal procedures or practices and the measures to rectify this.

23.3.2 Research Ethics Assessment of Confidentiality and Transparency in Less Serious Cases

From a research ethical point of view, cases that deal with less serious violations of research ethics norms should also be treated confidentially, at least until a decision has been made in the case.

Minor breaches of norms generally have no or little well-founded public interest, and active publication of such cases, or parts of them, are therefore rarely relevant. Serious journalists will probably not exercise their right of access to public documents or spend time and space on less serious matters that do not have broad public interest. However, some cases can be used in a rewritten or anonymised form as examples for training in research ethics.

23.3.3 Consideration of the Research Organization's Reputation

Consideration of the research organization's reputation is another factor to which many research leaders probably attach quite a lot of weight (without necessarily saying so) when deciding whether and how a case should be made public. Some then choose to keep the case as far away from public view as possible, even when there are conclusions that serious violations of research ethics norms have been committed. This is often a short-sighted strategy. An investigation into research misconduct usually involves many people, and many close colleagues in the research environment will understand that something is wrong. This often opens up for speculation, which can have a negative effect on the work environment. Leaks to the media must also always be expected. Few things are then more destructive to a research organization's reputation than an announcement that 'institute X is keeping research misconduct secret'. Research organizations are not harmed by an employee doing something wrong (it happens in all organizations), but they are judged on how they react, clear the matter up and the extent to which they are open about the wrongdoing. The best strategy from the organization's point of view is therefore being honest and open about what has happened, while at the same time protecting its employees with factual arguments. This will be respected and strengthens the reputation.

23.4 Special Duties and Roles of Those Involved in the Processing of the Case

23.4.1 The Suspect's and Witnesses' Duty

Researchers and others who are involved in or may have information in a case where there is a suspicion of a breach of research ethics norms should help to gather relevant documentation that may be used as evidence in the case. To refuse to do so or deliberately remove or destroy material in order to hide evidence may in the worst case be regarded as a serious violations of research ethics norms. This also applies to suspect(s) in the case.

23.4.2 The Management's Role

In cases of research misconduct and in the event of a crime, there is a general expectation that the top manager of the research organization, who has the overall responsibility, face the case both internally and externally. The top manager's and the other involved leaders' handling of the situation is therefore very important. These cases are often sensitive and complicated, and it is easy to make mistakes, not least because serious cases are so rare that management and staff may have little experience with them. This probably contributes to the fact that poor or incorrect case processing is not uncommon.

Many cases are also rooted in or develop into conflicts between researchers, or between a researcher and the management. These conflicts often escalate during the case processing, and the pile of documents that fail to shed any light on the research ethics of the case can grow to ominous heights. Three ways in particular of handling these cases can contribute to this happening:

- Some leaders take a premature stance on a whistleblower's allegations or the suspect's explanations. A whistleblower and the suspect will quickly notice this and often interpret it as the management being biased. A management that is not perceived as neutral, factual and solution-oriented, creates conflict of its own accord and can also cause the conflict to spread throughout the research environment around the directly involved parties.
- Some leaders let the whistleblower and the suspect play the lead roles in disagreements between the two and view their own role as that of 'judge' or 'mediator'. If the whistleblower and the suspect are colleagues, in such cases the management often creates a personal conflict or contributes to an underlying personal conflict developing further.
- Some managers allow consideration for their own and the research organization's reputation to influence the processing of the case. Serious cases almost always attract negative public attention and media coverage. When this happens,

there is nothing wrong with management seeking to protect the organization's reputation in an accountable manner. Some may, however, go too far by hiding or toning down the wrongdoing, modifying the investigation so that it becomes as 'kind' as possible, etc. Others may go the opposite way, becoming so preoccupied with showing that the organization and management have zero tolerance for violations of research ethics norms that the suspects are treated in an unfairly strict way. Both ways violate research ethics norms such as accountability, objectivity, openness and fairness and can lead to internal conflicts in the research organization.

23.4.3   The Whistleblower's Role

Because the whistleblower's research ethics competence and overview of facts may be limited, and because the person in question may have an interest in the case, it is unwise to uncritically base the investigation on the whistleblower's suspicions. A report from a whistleblower should be seen as a *signal* to the research organization's management that someone in the organization may have acted ethically irresponsibly. The management's task will then be to make up their own opinion on whether there are grounds for suspicion against someone. It is the suspicion that *then* possibly arises, factually and objectively formulated by a case manager in the organization, that the suspect should first and foremost be confronted with. A whistleblower that is also personally affected by the case must nevertheless still be heard, but then in the capacity of being a party to the case. It is this procedure that is used as a basis in the following.

23.5   Processing of Various Categories of Irresponsible Research Practice

All research organizations have their own procedures for handling deviations from responsible research practice. The procedures are adapted to their activities. The tasks and assessments that must be made in each case will, however, be approximately the same everywhere. This chapter provides a description of how such cases are typically dealt with or can be dealt with in a research organization.

23.5.1   Handling Sloppiness and Negligence

Sloppiness and negligence are often uncovered by colleagues, managers, clients or peers responsible for quality control. Few will find it natural to report sloppiness and negligence in a formal way; they will usually report directly to a project manager or first-line manager. Reacting to sloppiness and negligence differs little from ordinary, professional leadership and supervisory tasks and should therefore generally be done without it becoming a 'case'. The most important thing for the leader will be to ensure that the errors are corrected to the necessary degree and that those who have been sloppy or negligent make improvements. Particularly serious or repeated cases of sloppiness and negligence can, on the other hand, be handled more strictly and may be treated in the same way as less serious violations of research ethics norms; see below.

23.5.2 Handling Less Serious Violations of Research Ethics Norms

When less serious violations of research ethics norms are discovered or suspected, some organizations choose to let the immediate superior of the suspect be responsible for investigating the case and possibly also making a decision in it. Others refer the case upwards in the organization and/or separate the responsibility for the investigation and the decision. Some may have other arrangements. Such violations of norms – even if they are less serious – are often processed as a 'case' and documented accordingly. However, these cases can vary significantly. Some are simple and can be treated properly in a simple way, others are complicated and controversial and must be handled with precision. In cases where it is obvious that research misconduct can be ruled out; where the facts are clear and indisputable; where the suspect admits to having acted wrongly; or where no personnel conflicts exist, it is natural to let the case manager investigate and possibly also make a decision. In other, more complex cases, it may be appropriate or necessary to submit the case to an ethics commission or the like (see Sect. 3.5) before the management makes a decision. Using an ethics commission for advice is especially relevant if it leads to a better-founded management decision. When a group of peers have assessed the actions of the suspect the management's decisions may also be easier to accept by the parties in the case and others in general. When it is relevant to obtain advice from an ethics commission, the case will typically be handled in much the same way as in suspicions of research misconduct (see next section), possibly somewhat simplified and adapted to the seriousness and complexity of the case.

23.5.3 Handling Serious Breaches of Research Ethics Norms (Research Misconduct)

When there is a suspicion of research misconduct, the case must be investigated thoroughly before the management can make its decision. This is often a resource-intensive process that can amount to several man-weeks in case-processing time, often spread over a few months. If the decision in the case is appealed or leads to conflict, it may take a year or more before the case is closed.

Such cases are also demanding because statutory provisions and research ethics guidelines must be followed in order for whistleblowers and suspects to be treated properly, fairly and equally. Because there are few such cases, many managers and case managers have rather limited experience in this. Clear internal procedures in the research organizations are therefore particularly important here.

The Steps in the Handling of Suspicions of Serious Violations of Research Ethics Norms

The work with these serious cases is normally divided into steps. The order and content of each step may vary somewhat with the circumstances of the case. However, leaving out many details, the typical main tasks and assessments at each step can be summarized in brief and with some comments as follows (the presentation here is based on the case starting with a report from a whistleblower):

Step 1. Receive the whistleblower report and clarify what it concerns (responsible: the case manager).

- Get organized.

  - The research organization's internal procedures will have provisions for appointing a person responsible for handling the case, hereinafter referred to as the *case manager* (often a leader in the organization). In some organizations, the case manager may also have the decision-making responsibility, but the presentation below is based on the case manager and the decision-maker not being the same person. The case manager will either clarify the facts of the case alone, or leave (parts of) the work to a case officer.
  - A log should be kept of all incidents in the proceedings (received and sent documents and emails, meetings, etc.), and a case archive must be established so that all documents in the case are collected and easily accessible to those entitled to access.

- Clarify what the whistleblower's report concerns.

  - Many suspicions of breaches of responsible research practice come from a whistleblower and are often presented in a vague way. Facts are often mixed with subjective opinions and serious issues with less serious ones. In order to ensure that the whistleblower's report is correctly understood, an oral interview should be conducted and documented with minutes. This should be done immediately after receiving the report.
  - If there are multiple suspects (for example, when the suspicion is linked to a scientific publication with more than one author), the actions of each individual suspect must be clarified separately and the question of culpability assessed *individually*. All suspects shall also have the opportunity to explain themselves and express themselves individually in the case. However, the description below is based on there being one suspect.

- Make a decision on whether the case should be treated confidentially and thereby exempt from public access, or not.

    - In the interests of the suspect and any other parties concerned, the case should be treated confidentially at least until a conclusion is reached (see Sect. 23.3).

- Inform the suspect.

    - The suspect must be quickly informed of the whistleblower report and receive a copy of the written material that the whistleblower has provided. This can be done through an interview, where the case manager informs the suspect about the report and the way it will be handled by the employer. The suspect is asked to comment on the report and after the meeting provide a preliminary written account to which any documentation that sheds light on the case is attached. At the same time, the suspect should be informed that a final and complete explanation is desired only after the research organization has formed an independent opinion on the suspicion. The suspect should also be informed about the possibility of seeking help and advice from any employee's unions or others. The meeting should be concluded with information on the next steps in the process. The meeting must be documented with minutes.

- Provide easily accessible information from other people or sources.

    - Persons who may have information in the case should be contacted at an early stage. Initially, to save time, this can be done by phone (written explanations can be obtained later). Similarly, publications, reports, dissertations, emails and other documents that appear to be part of the case must be obtained.

Step 2. Assess the veracity of the report and decide if and how the case shall be processed further (responsible: the case manager).

- Make an independent assessment and specify the suspicion.

    - On the basis of the information from the whistleblower, the suspect's preliminary explanation and other information and documentation, the case manager must select the information and data that relate to violations of research ethics norms, and *in his or her own words formulate any potential suspicion*.
    - Many whistleblower reports contain issues that do not relate to research ethics. Accusations of research misconduct, for example, are not uncommon in conflicts between colleagues. Such personnel conflicts bring in personnel managers, HR managers, employee's unions and lawyers. Few of them have the research ethics competence and/or neutrality required to deal with breaches of research ethics norms in a credible manner. The accusations or suspicions of breaches of responsible research practice should therefore be dealt with separately, while the personnel conflict, if

possible, should be dealt with afterwards when any breaches of research ethics norms have been confirmed or disproved.

- Select one of five options.

  - Based on the suspicion that has now been formed, the case manager must decide on one of five alternative ways forward:

    1. It is beyond any doubt that no one in the organization has violated any research ethics norms. The case then requires no further investigation (some whistleblower reports are nonsensical, have nothing to do with research ethics, are directed at the wrong person or research organization, etc.). In some cases, there may be other matters that the management must deal with further on the basis of the report, such as scientific or personal conflicts between the suspect, the whistleblower and others.
    2. It is beyond any doubt that no research misconduct has taken place, but less serious breaches of responsible research practice may have occurred. The case can then be processed in accordance with the procedures for such less serious wrongdoing. As discussed above, however, there may be reasons why a less serious breach of norms should also be investigated more thoroughly and assessed by an ethics commission. The proceedings may then possibly progress as described below.
    3. There is reason to believe that there may have been serious violations of research ethics norms (research misconduct). The case is processed further as described below and moves into a phase that is often referred to as an 'investigation'.
    4. The suspect admits to having violated research ethics norms at this stage, accepts guilt and is willing to cooperate in documenting what has been done wrong. According to the circumstances, the case is then handled in as simple a way as possible. Initiating an investigation and engaging an ethics commission is then often an unwise use of resources. However, one must expect to spend some time documenting what has been done wrong and making sure that everything comes to light.
    5. There is reason to believe that there may be serious breaches of laws, regulations, procedures or agreements but not violations of research ethics norms. The case is then processed further by the management at a high level in the research organization, possibly with legal assistance. This is not discussed further here.

- Clarify who is a party to the case.

  - Based on the information now available in the case, the case manager must consider who the case may directly affect, i.e. who may be parties to it.[1] The suspect is always a party to the case, but others may also be affected,

---

[1] Who is *legally* a party to a case, and the parties' rights, may be specified in the legislation. The case manager must, as a minimum, follow the legal provisions that may be relevant to the case.

such as individual who has been plagiarized. The whistleblower is not a party to the case in his or her capacity as a whistleblower, but may be a party for other reasons.[2]

- The parties shall be notified in writing of their party status and rights. This includes being kept informed about the ongoing investigation of the case, gaining access to documents, having the opportunity to explain themselves and comment on other parties' information, or appealing the process and the decision.

• Inform the whistleblower of how the case will be handled.

- The whistleblower should now be thanked in writing for having reported the matter, and be informed of how his or her report will be processed further and that the organization later will report the outcome of the case. If only parts of the whistleblower's suspicions or allegations are pursued, the whistleblower must also be informed of the reasons for this and be given the opportunity to provide a written comment to which the research organization will then assess and respond. The whistleblower should be given the opportunity to formally complain if dissatisfied with the research organisation's assessment and handling of the case. A contact person for the whistleblower should be provided. With this, the whistleblower's involvement in the case in his or her capacity as whistleblower is over.
- If the whistleblower is also a party to the case, the case manager must now inform the whistleblower about the rights each party in the case has to be heard.

Step 3. Document the facts of the case and engage the research organization's ethics commission (responsible: the case manager).

• Appoint an ethics commission (or contact the ethics commission, if the organization has a permanent one) and inform it about the case, in order for the commission members to set aside time for the work.

- It is common to ask the ethics commission for *advice* only (decisions in the case are usually made by the management in the organization). This is used as a basis in the following.
- Sufficient information should be provided so that the commission members can assess whether they should recuse themselves from the case.

• Assess whether the suspect should be investigated for matters other than those that have emerged in the whistleblower report.

---

[2]The whistleblower will be a party to the case if the decision in the case is directed at or directly affects him or her. Examples may be that the decision in the case will be significant in a possible personal conflict in which the whistleblower is involved, or that the whistleblower is directly harmed (for example, unjustly omitted as a co-author). In general, the threshold for being considered a party to a case is high.

- – If, for example, there are indications of falsification in one journal article, there may be a reason to suspect that the person in question has also cheated in other works. Often this must be investigated, and the case will grow considerably in scope. In the following, it is assumed that this is not the case.

- • Formulate a draft mandate for the ethics commission's investigation into the case.

  - – In the mandate for the ethics commission's investigation, the work should be limited to matters in which the research organization requests the commission's assessment (see step 2).
  - – On the basis of this, the commission should be asked to answer specific questions. What one usually wants reasoned *answers* to are:

    - ▪ Whether violations of recognized research ethics norms have occurred, and what these are (the objective element of the action).
    - ▪ Whether the researcher has acted with intent or gross negligence (the subjective element of the action).
    - ▪ Whether there are published works that should be corrected or withdrawn.
    - ▪ Whether there are 'system errors' in the research organization, for example, inferior or incomplete internal procedures, unacceptable or illegal practices, inadequate training in research ethics, etc.

  - – The mandate must also state the legal and ethical regulations the commission shall base their assessments on. It should also allow the commission to look for additional evidence, orally question the parties to the case (or others who can shed light on the case), consult experts and more.

- • Document the facts of the case.

  - – This will usually include obtaining:

    - ▪ A final, comprehensive and written explanation from the suspect.
    - ▪ Written explanations from any other parties in the case.
    - ▪ Written explanations from others who can shed light on the case.
    - ▪ All documents relevant to the case.
    - ▪ Relevant internal procedures and other relevant information.
    - ▪ Any relevant research journals and supporting material.

  - – It is important that the suspect and any other parties in the case *are given the opportunity to comment on the specific issues that the ethics commission is to investigate* (this may, as discussed above, deviate from the whistleblower's suspicions or allegations). The parties must therefore be sent the case presentation and the ethics commission's mandate with an invitation to comment on this and to provide a uniform and complete explanation. If the parties believe that their previous, preliminary explanations are sufficiently comprehensive, they obviously do not have to submit new ones, but it is practical that the ethics commission has only one complete explanation from each party to relate to.

- Prepare the case presentation for the ethics commission.

  - The case presentation shall be the factual basis for the ethics commission's work. It must therefore be formulated as objectively and accountably as possible and provide an overall presentation of the case with document attachments, without the suspicions being assessed in reality.
  - In order to facilitate the ethics commission's work, it is appropriate that the case presentation only contains information and documents that the commission needs to assess the suspicion of violations of research ethics norms. Many case presentations overflow with documents in connection with the case processing in the research organization, layers upon layers of explanations and comments from whistleblowers and suspects, contributions from the parties' lawyers and employee unions, etc. By omitting material that is not relevant to the ethics commission's assessments, the commission's work will take less time, and the probability of incorrect assessments being made will be reduced. However, a complete event log and document list should be included in the appendices to the case presentation.

- Send the case presentation to:

  - The ethics commission.
  - The parties to the case, who shall have the opportunity to comment on the document (any comments should be forwarded to the ethics commission).
  - Top management in the research organization (for information).

Step 4. Investigate the case (responsible: the ethics commission).

- Review the case documents.

  - The commission's first task will be to familiarize themselves with the case and find out if it needs supplementary information or help from experts to clarify special professional or other issues for which the commission itself does not have competence.

- Interview the suspect in the case and, if necessary, other parties to the case.

  - The ethics commission should summon the suspect and any other parties in the case to an separate oral hearings. When the questioning is over, the person who has been interviewed should be given the opportunity more freely to explain matters that have not been touched upon or which are particularly important to bring up. Such face-to-face meetings are important to ensure that justice is done to the suspect and other parties, and is often clarifying for the commission when the question of guilt is to be considered. The invitation to the meeting must be in writing, state the purpose of the meeting, and provide information that the person to be interviewed may bring along a lawyer, a representative of an employee union or another. It is practical to make audio recordings of the meeting as a basis for written minutes that the interviewees must approve.

- If necessary, obtain statements from experts outside the commission.
  - In order to assess the severity of a breach of responsible research practice, it is often necessary to be an expert on the subject and research area in question. The commission can then seek help from external experts. Exceptionally, it may also be relevant to consult legal and research ethics expertise.
- Assess whether there are system errors in the research organization.
  - If the work on the case reveals something that may indicate a system error in the research organization, the commission must, if necessary, clarify the situation in more detail through an oral questioning of representatives of the management and/or written statements from them.
- Prepare the investigation statement.
  - The commission's statement should initially state the mandate for the investigation, provide an overview of the case with references to relevant documentation, and explain the commission's impartiality, method and the like. The commission's assessments and reasoned conclusions should then be discussed. The statement must document that the commission's work has been accountable and procedurally correct, thorough and accurate, objective, balanced, considerate and fair.
- Submit the report to the research organization.

Step 5. Make a recommendation for a decision in the case (responsible: the case manager).

- Send the ethics commission's statement to the parties to the case and obtain their written views on it.
  - Everyone who is a party to the case must be given the opportunity to comment on, respond to or point out what they believe is wrong or missing in the ethics commission's statement. The feedback should be given to the case manager.
- Make a recommendation in the case to the top management.
  - Unless decisive errors are discovered in the ethics commission's statements, or new aspects of the case emerge in the final phase, the commission's conclusions should be used as a basis for the case manager's recommendation in the case.
  - The case manager is responsible for preparing the final recommendation that will form the basis for the organization's decision on the case. The statements of the commission must always be attached.

- Send the final recommendation with attachments to:
  - The decision-maker in the research organization.
  - The parties to the case. They should be informed that comments on the final recommendation should be addressed to the decision-maker. They should be given a short but reasonable time to do so.
  - Other managers in the research organization that may be affected by the case (for information).

Step 6. Make a decision in the case and bring it to a conclusion (responsible: the decision-maker and the case manager).

- Take a decision in the case.
- Inform the parties to the case about the decision.

  - Their right of appeal must be stated. The right of appeal will depend on the research organization's internal procedures and the legislation to which the organization is subject.

- Inform others who need to know about the case.
- Inform the whistleblower about the outcome of the case.
- Consider informing the public about the case. If required by law, open the case to public oversight (see Sect. 23.3).

The case is now concluded. The suspect has either been found guilty or been acquitted of research misconduct. It then remains for the responsible leader to make a decision on any *reactions and sanctions* as a result of the case:

- Sanctions against anyone who has violated research ethics norms.
- Measures to correct and limit any damage caused by the misconduct.
- Measures that might improve or correct internal procedures and practices in the research organization, especially if the ethics commission has pointed out system errors.

The first two bullet points are discussed in more detail in Sect. 24.1.

23.6 Handling Ethically Reprehensible Matters in Activities That Are Not R&D

In addition to R&D, most research organizations are also engaged in other activities that may arise from or be associated with research, such as teaching, consultancy, laboratory services, engineering services, etc. Both researchers and other personnel may be involved in this. As discussed in Sect. 2.4.5, such activities should be carried out on the basis of the same professional ethical guidelines as for research activities, possibly with certain adaptations if the circumstances of the individual project so require. Cheating in connection with such activities is no less serious than cheating in research – and the research organization must respond to suspicions and investigate cases as consistently and conscientiously as in research. However, the legislation and ethical guidelines that are relevant for research usually do not apply to other activities. The research organizations must therefore have their own procedures for handling such cases.

Reference

All European Academies. (2017). *The European code of conduct for research integrity* (revised ed). ALLEA. Retrieved September 21, 2021, from https://allea.org/code-of-conduct/

Chapter 24
Reactions to Researchers Who Violate Research Ethics Norms

24.1 Disciplinary Actions from the Employer

When a research organization has concluded that an employee has violated research ethics norms, it must impose:

- Disciplinary action against the person who has acted wrongly, adapted to the severity of the case and the degree of culpability.
- Measures to prevent a recurrence, correct errors, limit any damage, and make sure justice is served for affected parties, etc.

The basis for this is the right to govern that an employer has towards its own employees. The organization's room for manoeuvre is limited by laws, internal regulations and possibly by employment contracts and any agreements with employee unions. When it comes to strict sanctions, one therefore quickly moves into the domain of the personnel/HR management and labour law lawyers.[1] This is beyond research ethics, and in the following, only some of a research organizations' main opportunities for disciplinary actions against its employees are described.

Research organizations are responsible for censuring *all* violations of irresponsible research practice, from negligence to research misconduct and more serious offences. This requires a series of forms of disciplinary censure, adapted to the severity of the case. Much is regulated by national laws that differ from country to

---

[1] Experience shows that some researchers who have been found guilty of wrongdoing do not accept the assessment that they have done wrong and/or oppose the measures the employer and others impose. In the worst case, this leads to the dispute being pursued by lawyers on both sides and possibly ending up in a courtroom. Both in order to do justice to those involved and for the employer to be able to exercise the right to govern in an efficient manner, the employer should therefore handle the cases in as correct a manner as possible. Specialists in handling these issues should, for example, be involved in the most serious cases.

© The Author(s), under exclusive license to Springer Nature Switzerland AG 2023
D. Slotfeldt-Ellingsen, *Professional Ethics for Research and Development Activities*, https://doi.org/10.1007/978-3-031-25484-0_24

country. However, some disciplinary measures are commonly available in many countries:

- Informal conversation with the nearest manager, project manager, supervisor or similar.

  – In minor and less serious cases of sloppiness, negligence or unacceptable practice, the most natural form of reaction is that the nearest manager, project manager or supervisor (for students, new employees, etc.) has an informal conversation with the person who has acted irresponsibly. The purpose is to:

    - Note that something has been done wrong.
    - Inform the employee about what is considered good practice, and make sure that this is understood.
    - Discuss and agree on measures to correct what has been done wrong, if relevant.

  – Unless it is an insignificant matter, a summary of the meeting should be made. The minutes should then be approved by the employee and employer.

- Formal conversation with the immediate manager.

  – In the case of more extensive, repeated or serious cases of sloppiness, negligence or unacceptable practice, the most natural form of reaction is that the researcher who has acted irresponsibly is summoned in writing to a formal conversation with their immediate manager. The purpose is to:

    - Inform the employee about what the research organization considers worthy of criticism.
    - Ensure that the employee understands and accepts the criticism. If this does not happen, it is natural to follow up the conversation with a written warning; see below.
    - Discuss and agree on measures to correct what has been done wrong, and prevent a recurrence. If no agreement is reached, the employer can use their right to govern to impose such measures.

  – The summons must state what the meeting is about and inform the employee of his or her rights to meet with support from an employee's union or others (employee's rights differ from country to country).

  – Minutes of the meeting should be made and approved by the employee and employer.

- Warning notice.

  – A warning is a strict form of reprimand that can in principle be used for all forms of seriously irregular behaviour in the workplace. A warning can be given orally, but should ideally be issued in writing so that the case is well documented and the chance of misunderstanding is reduced. Written warnings are often given as a follow-up to a prior conversation between the nearest manager and the person who has behaved wrongly. Warning or dismissal (see later) is a natural form of reaction in cases of research misconduct and serious breaches of employment contracts, assignment agreements and laws. These cases will always be thoroughly investigated by the research organization, and the parties to the case will have had the opportunity to comment along the way and on the conclusion (see Sect. 23.5). The purpose is to:

- Inform the employee that he or she is expected to adhere to responsible research practice in the future (along with an eventual specification of this), and give a warning about possible consequences if this does not happen (for example, dismissal).

  – It must be explicitly stated that a 'warning' is given. The idea behind a warning is to put pressure on employees to improve. If this becomes an isolated case, the person in question is only burdened with a note in the personnel file. If, on the other hand, the wrongdoer commits new, serious breaches of responsible research practice (or other wrongdoings), the warning is an important element in a possible subsequent dismissal case.

- 'Punitive' measures.
  In the event of particularly serious violations of research ethics norms, the research organization may react with various forms of what most people will perceive as a punishment (in reality a disciplinary measure, not be confused with punishment under the Penal Code, which only the courts can impose). These measures are subject to national legislation that vary by country. The most common sanctions in this category are:

  – Loss of seniority, demotion, relocation, forced termination of doctorial work and more.
  – Dismissal (alternatively discharge or suspension)

    - Dismissal implies leaving the job immediately, while discharge implies leaving at the end of the notice period that applies to the position. Suspension entails temporary removal from the workplace by order of the employer, usually used when it is appropriate for a suspect of wrongdoing to stay away during the investigation of the case. In order to make use of these disciplinary measures, the employer has to provide well-justified reasoning established in law.
    - Dismissal is an appropriate sanction in particularly serious cases of research misconduct. As in working life in general, dismissal may also be a relevant form of sanction for other serious breaches of trust and offences, for improper conduct in and outside the workplace, and for refusal to follow the employer's orders.

- Controlling, corrective and preventive measures.

  – In addition to the sanctions listed above, it may be necessary for the employer to ensure that everything wrong has come to light and that no new cases of wrongdoing occur. It may also be necessary to restore the trust between the research organization and others who may have been affected by the misconduct. Examples of such measures are:

- Review of other research works carried out by the guilty party in the case to ensure that everything has come to light (several cases show that those who cheat do so several times). However, this is often done at an early stage of the investigation of the case.
- Order or offer of training in research ethics and responsible research practice.
- Order that future research work shall be subject to special supervision or quality assurance.
- Deprivation of tasks and responsibilities in connection with teaching and supervision, management (including project management), project participation, committee activities, conferences and meetings, etc.
- Reallocation of internal and external (if possible) research funds.

- Measures such as this are particularly relevant when the guilty parties do not acknowledge or understand that they have acted incorrectly and when there is uncertainty about their willingness or ability to improve.

- Restorative measures

  - When someone acts irresponsibly, it is important to correct what may have been done wrong (for example, withdraw or correct errors in scientific articles). The researchers in question are essentially responsible for this. In serious cases of misconduct, however, it may be more natural for the research organization to take responsibility for clearing up inaccuracies, in collaboration with those involved. This is due to the fact that in most cases the research organization has legal responsibility for the research. The research organization, for example, is most often the contract partner for funding bodies and clients. Both the research community and the general public will also expect an employer to take responsibility for this. Relevant measures to reduce the damage of serious breaches of responsible research practice may, for example, include:

    - Correction or withdrawal of journal articles, reports, dissertations, information materials, websites, blogs, CVs, etc.
    - To provide information about the case to co-authors, partners, funding sources, clients, government agencies, publishers and others who are or may be affected by what has happened. The information should include how the case has been handled, and the measures that are taken in connection with the case.
    - To provide information to the public in general through conventional media channels or social media, whenever it is considered appropriate or necessary that the public becomes aware of the case. This was discussed in more detail in Sect. 23.3.1.

When a researcher acts in a blameworthy way, the research organization may also be harmed. There may be a loss of reputation if the case is not handled correctly and efficiently, or if the organization is also somewhat to blame. The research organization may also lose assignments, funding and collaboration as a direct or indirect consequence of the actions of its employees. Many of the measures discussed above can therefore be justified on the basis of the research organization's need to protect itself.

24.2 Sanctions from Others Than the Employer

In addition to the employer, there may be reactions from many others, such as:

- *Journals and publishers* may demand articles to be withdrawn or corrected. They may also refuse to publish future articles.
- *Clients and funding sources* can stop the project, demand repayment and any compensation. They may also demand the wrongdoers be removed from the project. In addition, clients and funding sources may refuse to provide future support for projects in which the wrongdoers participate.
- *Collaborators* may terminate the collaboration or demand wrongdoers be removed from the collaboration. Potential, future collaborators may also be reluctant to collaborate with the wrongdoers.
- *Educational institutions* may withdraw awarded academic degrees if scientific irregularities are found in master's theses or PhD dissertations.
- *Employees and colleagues* who have been harmed (for example, who have been plagiarized, left out as co-authors, etc.) may refuse to cooperate with the person who has harmed them.
- *Students* may require another supervisor.

These sanctions will normally be based on assessments made by those who demand or implement them, but here, too, laws and regulations, internal procedures and agreements may limit the scope of sanctions. Some of these sanctions must be justified and can in principle be disputed; others require no justification.

Research organizations should, as far as possible, take into account the scope for the type of sanctions listed in the examples above when deciding which measures the organization itself should implement.

24.3 Reactions from Public Authorities

Researchers who break public laws and regulations can, like everyone else, be investigated and be issued penalties or other sanctions in accordance with provisions that apply in each individual case. This may come in addition to sanctions that the employer and others have implemented (in criminal cases, the courts will probably take this into account when sentencing). In special cases, other public authorities may also impose sanctions on the basis of their own authority and their own assessments.

Chapter 25
Possibilities for Reinstatement

25.1 Preconditions for Reinstatement

A person who is dismissed or deprived of duties and responsibilities as a result of research misconduct, will in some cases want to 'come back', i.e. become a researcher again or get back tasks and responsibilities, at least after some time. The management of the research organization that must take a position on this, then faces classic, moral dilemmas: When should one forgive? How can one be fair both to the person who has acted wrongly and to those who are harmed by the act? To make the right choice, one must think logically and rely on formal rules and common practice. Two factors will be central to the assessment:

The first has to do with *qualifications*: To be qualified for a position means having the knowledge, skills and attitudes that the position requires. To be qualified for a research position, scientific knowledge and merits are then not enough. One must also have good research morals. Research misconduct and other serious and deliberate breaches of responsible research practice must be seen as proof that the wrongdoer does not have the moral attitudes that a research position requires. A first condition for researchers to be reinstated is therefore that they have changed their moral attitudes in a credible way.

The second has to do with *the interests of the research organization, colleagues, partners, clients and others*. In many cases, it can be difficult and harmful for the research organization to give researchers who have violated research ethics norms a second chance. Institutes characterized by commissioned R&D will be particularly vulnerable in that the clients can refuse to have anything to do with the researcher, meaning that the economic basis for the researcher's position disappears. If the violation of responsible research practice has affected colleagues and partners, the researcher's presence can seriously damage the working environment. Another prerequisite for being reinstated as a researcher is therefore that it is not unreasonably detrimental to the research organization and others affected.

25.2   Reinstatement Following Loss of Responsibility and Work Tasks or After Reassignment

In many countries, it is common to expect that the employer will give employees who have done something wrong a chance to improve. This is one of several goals of the corrective and preventive measures described in Chap. 24. Being deprived of tasks and responsibilities will therefore often be seen as a measure subject to a time limit. Those who have acted irresponsibly are given a chance to show willingness and an ability to improve. An open and honest admission of wrongdoing, and credible assurances of having learned from it, will be a first step in the right direction. Anything to the contrary will be worrying. Concrete actions such as following courses and lectures in research ethics, choosing a mentor to give guidance, actively participating in activities that raise awareness of how ethics are integrated into science, etc. can make it more likely that the researcher's moral attitudes are showing positive development. Unfortunately, experience shows that some researchers also develop in a negative direction and, rather than improving, go to war against colleagues, partners or management. This often leads to the research organization being left with a personnel problem which it must solve with the instruments provided for in laws, internal procedures and employment contracts.

25.3   Reinstatement Following Dismissal

A research organization has neither a legal nor a moral duty to reinstate a dismissed researcher. Like everyone else, the dismissed researcher can of course apply for any vacant research positions, but applying for a position in the organization from which he or she was fired shows little respect for the problems caused there. However, applying for a position in another research organization cannot be considered wrong, at least not after a certain period of time. The researcher then has to compete with others for the position in the usual way. In addition, it will be crucial that the researcher is open and honest about his or her past and can prove to have the professional ethical knowledge and attitudes that the position requires. The research organization, for its part, must assess whether the applicant, despite his or her past, really can function well in the position without being to a detriment to the research organization and others.

The manufacturer's authorised representative in the EU is Springer
Nature Customer Service Centre GmbH, Europaplatz 3, 69115 Heidelberg,
Germany. If you have any concerns regarding our products, please
contact ProductSafety@springernature.com

Printed and bound by CPI Group (UK) Ltd, Croydon, CR0 4YY
29/04/2026
02099466-0004